Lowell E. Moser

Roots, Nutrient and Water Influx, and Plant Growth

ASA Special Publication Number 49

Proceedings of a symposium
sponsored by Divisions S-4, S-2,
and C-2 of the Soil Science Society of America,
Crop Science Society of America, and the
American Society of Agronomy
in Anaheim, CA, 28 Nov.–3 Dec. 1982.

AF895023

Editorial Committee
S.A. Barber
D.R. Bouldin

Managing Editor
David M. Kral

Assistant Editor
Sherri L. Hawkins

1984
Published by the
SOIL SCIENCE SOCIETY OF AMERICA
CROP SCIENCE SOCIETY OF AMERICA
AMERICAN SOCIETY OF AGRONOMY
677 South Segoe Road
Madison, WI 53711

Cover Design: Julia M. Whitty

Copyright 1984 by the Soil Science Society of America, Crop Science Society of America, and American Society of Agronomy, Inc.
ALL RIGHTS RESERVED UNDER THE U.S. COPYRIGHT LAW OF 1978 (P.L. 94-553). Any and all uses beyond the limitations of the "fair use" provision of the law require written permission from the publisher(s) and/or the author(s); not applicable to contributions prepared by officers or employees of the U.S. Government as part of their official duties.

Soil Science Society of America
Crop Science Society of America
American Society of Agronomy
677 South Segoe Road, Madison, WI 53711 USA

Library of Congress Catalog Card Number: 84-51525.
Standard Book Number: 0-89118-082-6.

Printed in the United States of America

Table of Contents

Foreword

Plant growth depends on an intricate interaction of soil factors, atmospheric factors, and plant genotype. An understanding of this soil-plant-atmosphere complex is necessary if the yields of field, horticultural, and forest plants are to be increased. The aerial portion of plants has received most of the attention of researchers over the years, while the root system has received considerably less focus.

This publication describes some recent studies on plant roots. The eight authors have presented a broad coverage of the subject from basic mechanisms to an overview of the role of roots in determining crop yields.

The American Society of Agronomy, the Crop Science Society of America, and the Soil Science Society of America are pleased to have sponsored the symposium and this resultant publication entitled *Roots, Nutrient and Water Influx, and Plant Growth.*

K. J. Frey, President
American Society of Agronomy

W. F. Keim, President
Crop Science Society of America

D. R. Nielsen, President
Soil Science Society of America

Preface

Plant and soil science meet at the root-soil interface. In recent years, we have recognized that root and soil interact at this interface so that the soil solution composition at the root-soil interface is different from bulk soil solution and hence influences the interpretation of soil chemical studies. This effect is influenced by nutrient influx kinetics and the size and morphology of the root system. Within the last 30 years, progress has been made toward understanding these complex soil-root-crop interactions. This publication is, firstly, a clear recognition of the importance of the subject and, secondly, a summary of the present state of knowledge. The 1982 program chairman for Div. S-4, Soil Fertility and Plant Nutrition, D. L. Grunes, asked S. A. Barber to organize a symposium on this subject for the annual meeting held in Anaheim, CA in 1982. The symposium was cosponsored by Div. C-2 and S-4 of the American Society of Agronomy. Six scientists, actively doing research with plant roots, in different areas were invited to give 30-min presentations and to prepare manuscripts for publication. A large attendance at the symposium indicated a significant interest in soil-plant interactions.

This publication, *Roots, Nutrient and Water Influx, and Plant Growth*, is the written version of the papers presented at the symposium. The first chapter by André Läuchli discusses our current knowledge on ion absorption mechanisms of plant roots. The second chapter by A. J. M. Smucker develops the recent knowledge on C flow from the shoot to the root, the use of the C by the root and C exudate from the root. The third chapter by M. G. Huck treats water absorption kinetics for roots growing in soil. In the fourth chapter, S. A. Barber and M. Silberbush discuss root morphology and use a mechanistic model to evaluate its significance. Then, P. H. Nye discusses the role of the root in altering soil properties so that they affect ion uptake. In the final chapter, D. A. Brown and H. D. Scott give an overview of the significance of plant roots in determining crop yield.

We are grateful for the conscientious efforts of the authors in preparing their chapters. The chapters will be important for stimulating interdisciplinary cooperation among soil and plant scientists in an area that will have increasing significance.

Editors
Stanley A. Barber
David R. Bouldin

Chapter 1

Mechanisms of Nutrient Fluxes at Membranes of the Root Surface and their Regulation in the Whole Plant[1]

ANDRÉ LÄUCHLI[2]

Roots are the sites of uptake by higher plants of mineral nutrient and other ions from the soil. A part of the ions taken up accumulate in root cells, particularly in their vacuoles, and ions can also be transported through the root to the xylem and upwards to the shoot. There is also some ion movement in the opposite longitudinal direction, that is downwards to the root apex. Finally, ions can be lost from the root by efflux from the outer cells to the soil. These ion transport processes are governed by a number of membrane transport steps, both at the cells of the root surface as well as the stele.

This chapter addresses two questions. First, the main mechanisms of ion fluxes at the plasmalemma and tonoplast of root surface cells will be discussed. Second, an attempt will be made to integrate these ion fluxes in the whole plant in terms of regulation of fluxes and shoot-root interactions. Figure 1 shows a model of the pathways of ion transport in roots, depicting influx across the plasmalemma at the root surface followed by vacuolar accumulation or symplasmic transport to the stele and ion release to the xylem vessels. Note that the apoplasmic (cell

---

[1]Contribution from the Univ. of California, Davis, CA 95616.
[2]Professor of plant nutrition, Dep. of Land, Air, and Water Resources, Univ. of California, Davis, CA 95616.

Copyright © 1984 Soil Science Society of America, Crop Science Society of America, and American Society of Agronomy, 677 South Segoe Road, Madison, WI 53711. *Roots, Nutrient and Water Influx, and Plant Growth.*

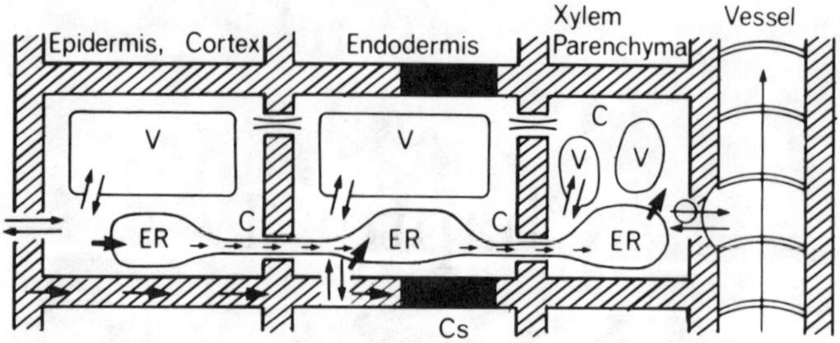

Fig. 1. Pathways of ion transport in roots. Note the endoplasmic reticulum (ER) as a structural component of the symplasmic pathway. Cs = Casparian strip (Läuchli, 1976b).

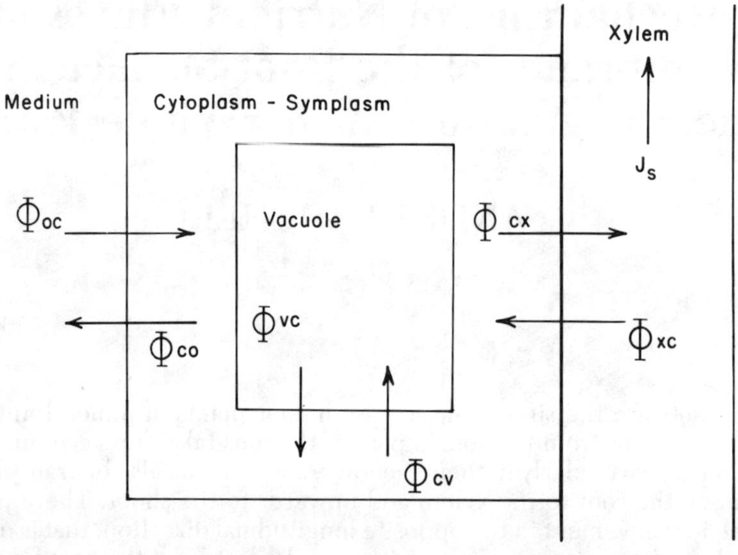

Fig. 2. Model of principal membrane fluxes in a root (net salt flux in xylem, $J_s = \Phi_{cx} - \Phi_{xc}$).

wall) pathway between cortex and stele is blocked by the Casparian strip (Cs) in the radial walls of the endodermis. The endoplasmic reticulum (ER) and the plasmodesmata are envisioned to form an important structural component of the symplasmic pathway. Symplasmic and apoplasmic transport have been reviewed extensively by Spanswick (1976) and by Läuchli (1976a), respectively. The principal membrane fluxes which govern ion transport in roots are exhibited in Fig. 2. The three membranes shown in the figure are the plasmalemma (influx to cytoplasm Φ_{oc} and efflux Φ_{co} and tonoplast (influx to vacuole Φ_{cv} and efflux Φ_{vc}) of the surface cells at the outer boundary of the root symplasm and the plasmalemma of the xylem parenchyma cells (efflux to xylem vessel Φ_{cx} and influx Φ_{xc}) at its inner boundary. The net ion uptake from the medium to the cytoplasm-symplasm is determined by influx and efflux and given by

$$\Phi_{net} = \Phi_{oc} - \Phi_{co} . \quad [1]$$

Ion uptake by plant roots has been treated comprehensively by Hodges (1973), Pitman (1976), Clarkson and Hanson (1980), Clarkson (1981), and Leonard (1984). Transport of ions across the root into the xylem has also been the subject of a number of recent reviews (Anderson, 1976; Läuchli, 1976b; Pitman, 1977, 1982a, 1982b; Bowling, 1981; Lüttge, 1983).

MECHANISMS OF NUTRIENT FLUXES AT CELL MEMBRANES OF THE ROOT SURFACE

Active and Passive Ion Fluxes

Root cell membranes have a low permeability for ions and show a high degree of selectivity in ion transport, for example selective transport of K^+ relative to Na^+ (Epstein, 1972). These well-known facts have led to the carrier concept of ion transport through root cell membranes (review: Epstein, 1976). Briefly, it was postulated that carrier-mediated transport of an ion across a cell membrane is analogous to the process of enzyme-mediated catalysis of a substrate (Epstein and Hagen, 1952) and follows Michaelis-Menten kinetics according to the equation

$$v = (V_{max}\ c_o)/(K_m + c_o)\ . \qquad [2]$$

v is the rate of transport, V_{max} the maximal transport rate, c_o the ion concentration in the medium, and K_m the Michaelis constant. This mechanism of transport operates at low external concentrations (e.g., < 1 m*M* K^+). At higher concentrations rates of transport rise above the theoretical value of V_{max} of the low concentration, high affinity mechanism. As described by Epstein initially, this second mechanism displays curvilinear or more complex kinetics (Epstein, 1976). The duality referred to above has been questioned (Nissen and Nissen, 1983). Recently, Kochian and Lucas (1982) reaffirmed the dual nature of the transport process. In their experiments with corn (*Zea Mays* L.) roots, the kinetics of K transport were resolved into a saturable component at low external concentrations (in agreement with Epstein's finding) and a linear component of unknown nature—but probably not due to simple diffusion kinetics—at higher concentrations. Pitman (1976) attempted to bring together the results of the carrier concept with measurements of fluxes and electrical potential differences (PD) between cells and the external medium. Consideration of fluxes and cell PD allows to distinguish between active and passive ion fluxes and to integrate ion carriers into the concepts of active transport. A detailed account of the biophysical considerations of ion transport has been presented by Lüttge and Higinbotham (1979).

If the permeability of the cell membrane for cations and anions is different (permeability coefficients $P_{C}+ \neq P_{A}-$), a PD is the consequence. This PD has the characteristics of a diffusion potential and can be described in general terms by the Nernst equation

$$E_j = (RT/z_jF)\ln(a_j^o/a_j^i) , \quad [3]$$

where E_j is the Nernst potential for an ion species j, R, and F are the gas and Faraday constants, respectively, T the absolute temperature, z_j the valence of the ion j, and a_j^o and a_j^i the ion activities outside and in the cell. When PD is measured between the vacuole of the cell and the external medium (E_{vo}), the values are usually −50 to −250 mV (inside negative). If an ion moves passively across a cell membrane, its distribution between inside and outside is determined by the electrochemical potential gradient, and its Nernst potential E_j is equal to the measured membrane potential E_M. However, for many ions E_j calculated according to Eq. [3] is very different from the measured E_M and, thus, *active ion transport* in one direction (against the electrochemical potential gradient) is indicated. In the case of active transport the ion distribution between inside and outside is determined by the electrochemical potential gradient and additional driving forces.

In studies of transport of monovalent ions, the relative permeabilities of the ions in the membrane are to be considered and the Nernst Eq. [3] is usually replaced by a modification of the Goldman equation (Lüttge and Higinbotham, 1979; Pitman, 1982a):

$$E = (RT/F)\ln\{([K_o^+] + \alpha[Na_o^+] + \beta[Cl_i^-])/([K_i^+] + \alpha[Na_i^+] + \beta[Cl_o^-])\}, \quad [4]$$

where bracketed terms are concentrations of the particular ions outside (o) and inside (i) and α and β are permeability ratios for Na^+/K^+ and Cl^-/K^+, respectively. Permeability ratios are derived from electrical and flux measurements (using radiotracers); for barley (*Hordeum vulgare* L.) roots values for α are about 0.2 to 0.3 and for β about 0.03 (Pitman, 1969). Similar to the Nernst equation, active transport is indicated when the diffusion potential E calculated from Eq. [4] markedly deviates from the measured membrane potential. Note that Eq. [4] is only applicable if K^+, Na^+, and Cl^- are the major moving ions; NO_3^-, an additional important anion is neglected in this equation.

Fluxes measured by means of radiotracers may also be used to distinguish between active and passive ion transport. The distinction is based on the Ussing-Teorell equation with the flux ratio for an ion j given by

$$\Phi_j^{oi}/\Phi_j^{io} = a_j^o/(a_j^i e^{z_j F E_M/RT}) , \quad [5]$$

E_M = measured membrane potential; other symbols explained above. If fluxes do not obey Eq. [5], it is concluded that active transport is occurring.

The thermodynamic criterion outlined above is frequently replaced by the dependence on energy-supplying reactions of metabolism as a criterion for active transport. Dependence on energy-supplying reactions of metabolism was the first measure of active transport, used by

Hoagland, Lundegårdh, Steward, and others (Epstein, 1972). This criterion was subsequently replaced by the thermodynamic one. The coupling of ion transport with metabolic energy can be experimentally demonstrated by the dependence on temperature or the effects of anaerobiosis and inhibitors. The use of inhibitors, however, is often fraught with problems, for inhibitors may unspecifically affect metabolism and, moreover, they may also influence membrane permeability to ions (Jackson, 1982) and the hydraulic conductivity of a root (Pitman et al., 1981b). To determine the temperature dependence of ion transport the temperature quotient Q_{10} is measured (change in transport rate v when temperature is lowered or increased by 10°C). (Temperatures close to the freezing point should be avoided, because the transient cold shock at such low temperatures observed by Chastain and Hanson, 1982 and Zocchi and Hanson, 1982 may drastically affect the results). Transport rates v_1 and v_2 at two temperatures t_1 and t_2 are given by

$$\ln Q_{10} = [10/(t_2 - t_1)] \ln(v_2/v_1) \quad [6]$$

(Lüttge and Higinbotham, 1979). From v at various temperatures, the activation energy μ_e of the transport process can be calculated using the equation

$$\mu_e = (Rt_1t_2)/(t_2 - t_1) \ln(v_2/v_1) \ . \quad [7]$$

Q_{10} values for diffusion in an aqueous medium are about 1.2 to 1.5 but much higher for metabolic reactions. Diffusion across membranes often leads to Q_{10} values considerably higher, and thus a high Q_{10} value is not an unequivocal criterion for the dependence of ion transport upon metabolic energy. Hence, the dependence upon metabolism of ion transport across plant cell membranes is not a rigorous criterion for active transport. Nonetheless, it can give insight into the energetic coupling of transport to biochemical reactions.

Model of Ion Fluxes at Plasmalemma and Tonoplast

Jeschke (1980a, 1980b) presented a model depicting the primary fluxes of monovalent cations and anions at the plasmalemma and tonoplast of outer root cells. This model is shown in Fig. 3. The central active mechanism at the plasmalemma appears to be a proton efflux pump which is energetically powered by adenosine 5′-triphosphate (ATP) (1 in Fig. 3). The H^+ efflux pump is electrically balanced by cation influx with pronounced K^+/Na^+ selectivity (2). The H^+ pump is an electrogenic pump, that is a pump which transports a net charge across the membrane (Poole, 1978; Lüttge and Higinbotham, 1979; Spanswick, 1981 for a more detailed discussion of electrogenic pumps); it generates an electrical potential Ψ given by the equation (Poole, 1978)

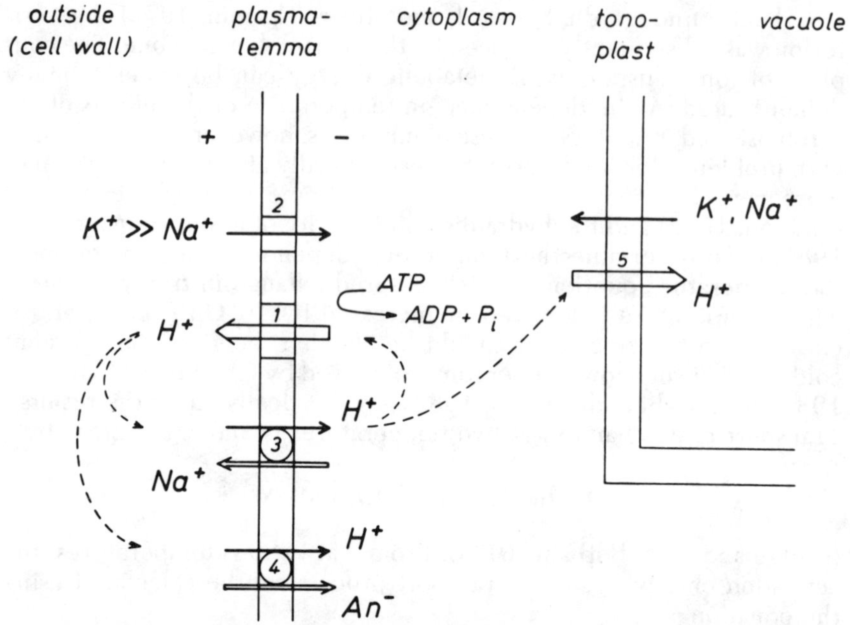

Fig. 3. Model of primary fluxes of monovalent cations and anions at plasmalemma and tonoplast of an outer root cell (Jeschke, 1980a).

$$\Psi = \Delta G_{ATP}/nF + 59\Delta pH \ , \qquad [8]$$

where ΔG_{ATP} is the change in free energy for the hydrolysis of 1 mole of ATP and n is the stoichiometry of the pump; n is considered to be $2H^+$/ATP (Poole, 1978). The H^+ pump drives Na^+ efflux (3 in Fig. 3) by H^+-Na^+ antiport and anion influx (4) by H^+-anion symport. (The term "antiport" relates to transport of two ions with identical charge in opposite directions, and the term "symport" denotes transport of a cation and an anion in the same direction; both transport processes involve the participation of membrane carriers.) In Jeschke's view, protons in the overall transport scheme cancel each other out and, as far as cation fluxes are concerned, there is K^+-Na^+ exchange. The situation at the tonoplast is less clear, but Jeschke also postulates an H^+ pump (H^+ influx pump, 5) which is electrically balanced by cation efflux with K^+/Na^+-selectivity.

Support for the postulated H^+ efflux pump at the plasmalemma is strong (Pitman, 1982a). Particularly, the use of the fungal toxin fusicoccin (review: Marré, 1979) has been very illuminating. Fusicoccin stimulates H^+ efflux from roots and also stimulates K^+ influx and hyperpolarizes the membrane potential. Pitman's and Marré's groups consider the evidence from the fusicoccin studies in support of the model presented in Fig. 3, where there is indirect (electrogenic) coupling between H^+ efflux and K^+ influx (e.g., Pitman et al., 1975; Bellando et al., 1979). On the other hand, Glass and Siddiqi (1982) question this model, on the basis of a detailed study of the stoichiometry of coupling between H^+ efflux and K^+ influx, and support the alternative view that K^+ influx is directly powered by ATP with the involvement of an ATPase (see section Energetics of Ion Fluxes). Whatever the nature of linkage between H^+ efflux and K^+ influx, Glass et al. (1981) demonstrated an excellent correlation between net H^+ efflux and K^+ influx in roots of 24 barley varieties (Fig. 4) and suggested that the determination of net H^+ efflux by means of the indicator methyl red be used as a method of screening crop genotypes for efficiency in K^+ uptake.

In Jeschke's model (Fig. 3), anion influx occurs via an H^+-anion symport. Formally, this is identical with an anion/OH^- antiport. It has long been known that when NO_3^- is taken up, roots are able to excrete OH^- (e.g., Ben Zioni et al., 1971). Recent evidence suggests that in corn roots in which the NO_3^- uptake system is induced by the presence of NO_3^- in the external medium, the apparent stoichiometry of NO_3^- influx to OH^- efflux is 2:1 (Thibaud and Grignon, 1981). Active phosphate influx into roots does not appear to be directly related to the H^+ efflux pump depicted in Fig. 3; rather, an electrogenic phosphate pump may operate at the plasmalemma of root cells, which appears to be driven by a pH gradient across the membrane, as indicated to operate in white clover (*Trifolium repens* L.) roots (Dunlop and Bowling, 1978). Another important active transport mechanism at the plasmalemma of root cells not exhibited in Fig. 3 is the recently postulated Ca^{2+} efflux pump (Macklon, 1975; Macklon and Sim, 1981; Dieter and Marmé, 1980). The ionic activity of Ca in the cytosol of the cytoplasm is probably about 10^{-6} *M* or less; this low Ca^{2+} activity is postulated to be

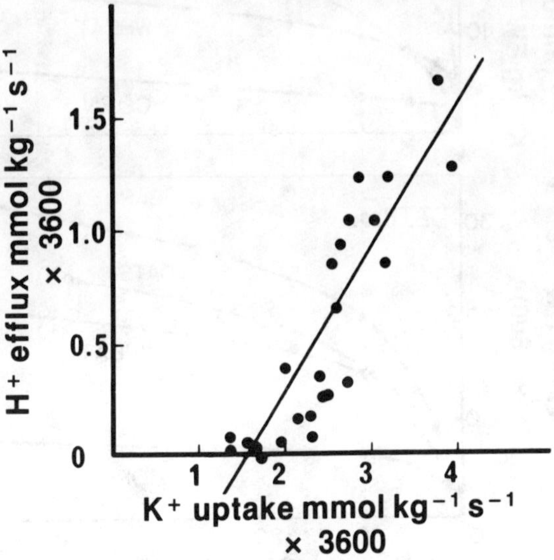

Fig. 4. Correlation between net H^+ efflux and K^+ influx in roots of 24 barley varieties. Roots of intact seedlings exposed to 1 m*M* K_2SO_4 plus 0.5 m*M* $CaSO_4$ for 24 h (r = 0.88) (Glass et al., 1981).

maintained by a series of active transport mechanisms, that is active Ca^{2+} efflux at the plasmalemma driven by a membrane-bound Ca^{2+}-ATPase and active Ca^{2+} influx into both vacuole and mitochondria (Marmé, 1983; Hanson, 1984).

Energetics of Ion Fluxes

Active and energy-dependent ion uptake requires the expenditure of metabolic energy supplied by aerobic respiration as ATP (e.g., Epstein, 1972; Poole, 1978). There is now overwhelming evidence that an ATPase (cleaving ATP to ADP + P_i) bound to the plasmalemma is the energy-transducing agent between ATP and active H^+ efflux and cation influx (Leonard, 1983, 1984). First experimental evidence came from the strong correlation between K^+ influx and K^+-stimulated ATPase activity of an isolated cell membrane fraction in a variety of cereal roots (Fischer et al., 1970; Fig. 5). Following this pioneering study, research mainly conducted in the laboratories of Hodges and Leonard led to the detailed biochemical characterization of the plasma membrane-ATPase and its role in cation influx, particularly K^+ influx in roots

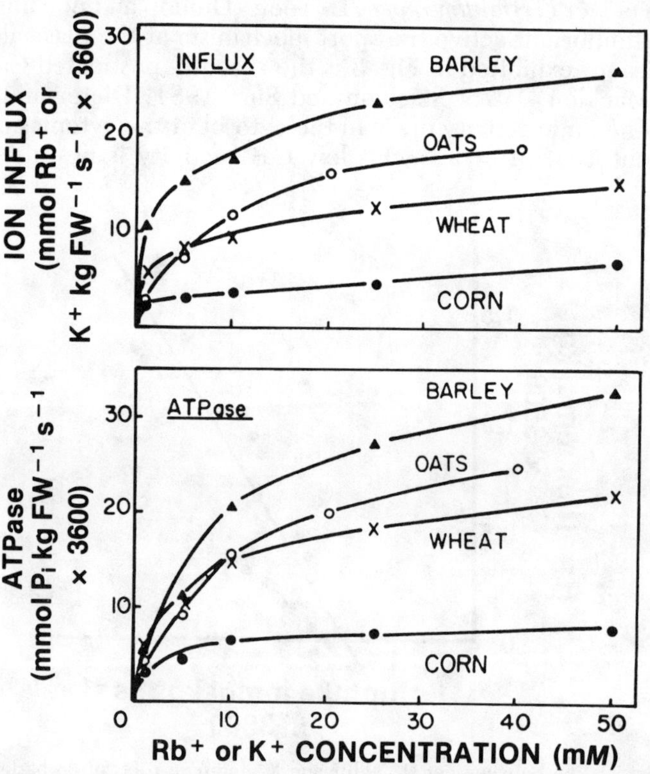

Fig. 5. Correlation between influx of K^+ and K^+-stimulated ATPase activity of membrane preparations in four different cereal roots (r = 0.94) (Fisher et al., 1970).

(reviews: Hodges, 1976; Läuchli and Pflüger, 1979; Leonard, 1983, 1984). This ATPase is likely to be bound to the plasmalemma, has a pH optimum of about 6.5, and probably uses $Mg^{2+}\cdot$ATP as the true substrate (Leonard, 1983). The latter fact explains the absolute Mg^{2+}-requirement of this enzyme. The activity of this ATPase is further stimulated by alkali chlorides (Table 1). Maximum enzyme activity occurs in the presence of KCl, and the monovalent cation stimulation of enzyme activity is progressively less when KCl is substituted by the alkali series RbCl to LiCl (Table 1). It is still debated whether the increment of enzyme activity referred to as K^+-stimulated ATPase activity is directly related to K^+ transport through the membrane. The good correlation in cation specificity and kinetics between alkali cation stimulation of ATPase activity and alkali cation influx in root cells is, however, in support of the relation of the enzyme to K^+ transport (Table 1, Sze and Hodges, 1977).

In studies of location and function of membrane ATPases it is important to differentiate between ATP phosphohydrolases from various locations in the cell and substrate-unspecific phosphohydrolases. Unfortunately, there is no known inhibitor which specifically inhibits both K^+-stimulated ATPase activity derived from the plasmalemma and K^+ influx into the root (Leonard, 1983). The use of several metabolic inhibitors, in combination with pH values that are optimum for the enzyme under study, however, provided a means of distinguishing the plasmalemma ATPase from other phosphohydrolases (Gallagher and Leonard, 1982; Table 2). The information compiled in Table 2 shows that the mitochondrial membrane ATPase is specifically inhibited by azide but the plasmalemma ATPase is not. The latter enzyme is inhibited by vanadate but so is the substrate-unspecific acid phosphatase; however, the differentiation between these two enzymes is possible when molybdate, an acid phosphatase inhibitor, and vanadate are used together. Furthermore, among the three inhibitors listed only vanadate inhibits plasmalemma ATPase activity and K^+ influx in roots. Vanadate

Table 1. Effect of monovalent cations on plasmalemma ATPase activity of roots (modified from Leonard, 1983).

	ATPase activity† (nmol P_i kg protein^{-1} s^{-1})	
Addition	Oat‡	Corn§
Mg^{2+}	8.37	3.73
Mg^{2+} + KCl	16.7 (100)¶	5.48 (100)
Mg^{2+} + RbCl	15.6 (87)	5.28 (89)
Mg^{2+} + NaCl	15.3 (83)	5.17 (83)
Mg^{2+} + CsCl	14.8 (78)	5.09 (78)
Mg^{2+} + LiCl	11.8 (42)	4.56 (47)

† Assay conditions: 3 m*M* ATP, pH 6.0–6.5,
additions: 3 m*M* Mg^{2+}
50 m*M* monovalent salt.

‡ From Leonard and Hodges, 1973.

§ From Leonard and Hotchkiss, 1976.

¶ Numbers in parentheses denote the increase in ATPase activity due to monovalent salt relative to KCl.

Table 2. The ATP phosphohydrolases in corn roots: sensitivity to inhibitors (after Gallagher and Leonard, 1982).

Inhibitor	Plasmalemma ATPase	Mitochondrial membrane ATPase	Acid phosphatase
pH:	6.4	9.0	5.0
Vanadate† (0.01–0.5 m*M*)	+	−	+
Molybdate (0.01–1 m*M*)	−	−	+
Azide (0.1–10 m*M*)	−	+	−

† Inhibits K^+ influx.

has also been demonstrated to inhibit strongly the stimulating effect of fusicoccin on H^+ efflux and K^+ influx (Colombo et al., 1981). Thus, the use of vanadate in combination with molybdate and azide allows one to ascertain whether one deals with a plasmalemma ATPase or another phosphohydrolase.

A model developed by Leonard (1983) on the possible structure and function of the plasmalemma ATPase in plant cells is depicted in Fig. 6. Briefly, the enzyme is viewed as a multi-subunit enzyme complex which is an intrinsic membrane protein. The large subunits form aqueous channels through the ATPase which contain the binding sites for H^+ and K^+. The catalytic site for Mg·ATP hydrolysis is located on the cytoplasmic side of the membrane-bound enzyme. Binding of H^+ to the large subunits on the cytoplasmic side induces Mg·ATP dependent phosphorylation of the subunits, causing a conformational change of them that allows H^+ to move through the aqueous channel to the external medium. Binding of K^+ to the large phosphorylated subunits on their outer side causes dephosphorylation, return to the original conformation of the large subunits, and K^+ transport through the aqueous channel to the cytoplasm. According to this view, the plasmalemma ATPase is an H^+-K^+ exchange pump which utilizes ATP. Thus, it would serve both as the H^+ efflux pump (Fig. 3) and the K^+ selective ion carrier (Epstein, 1976). Evidence obtained from sealed membrane vesicles is in support of the concept that the plasmalemma ATPase is an electrogenic H^+-K^+ exchange pump (Sze, 1980; Sze and Churchill, 1981). (Sealed membrane vesicles are vesicles derived from membranes and are not leaky but perform transport functions like membranes in intact cells.) The controversy as to whether this ATPase is primarily an H^+ efflux pump or a K^+-selective cation influx pump has been authoritatively reviewed by Leonard (1982).

Is there also a tonoplast-associated membrane ATPase involved in H^+ and K^+ transport? Jeschke (1980a, 1980b; Fig. 3) postulated an H^+ influx pump at the tonoplast which is electrically balanced by K^+ efflux from the vacuole. Tonoplast vesicles isolated from vacuoles of tobacco (*Nicotiana tabacum* L.) cell protoplasts did not have substrate specific ATPase activity (Briskin and Leonard, 1980), but more recent research indeed indicated the probable location of an ATPase at the tonoplast of root cells. Work on vacuoles isolated from the storage root of red beet (*Beta vulgaris* L.) produced clear evidence for the presence of an enzyme which was quite specific for Mg·ATP, and the activity

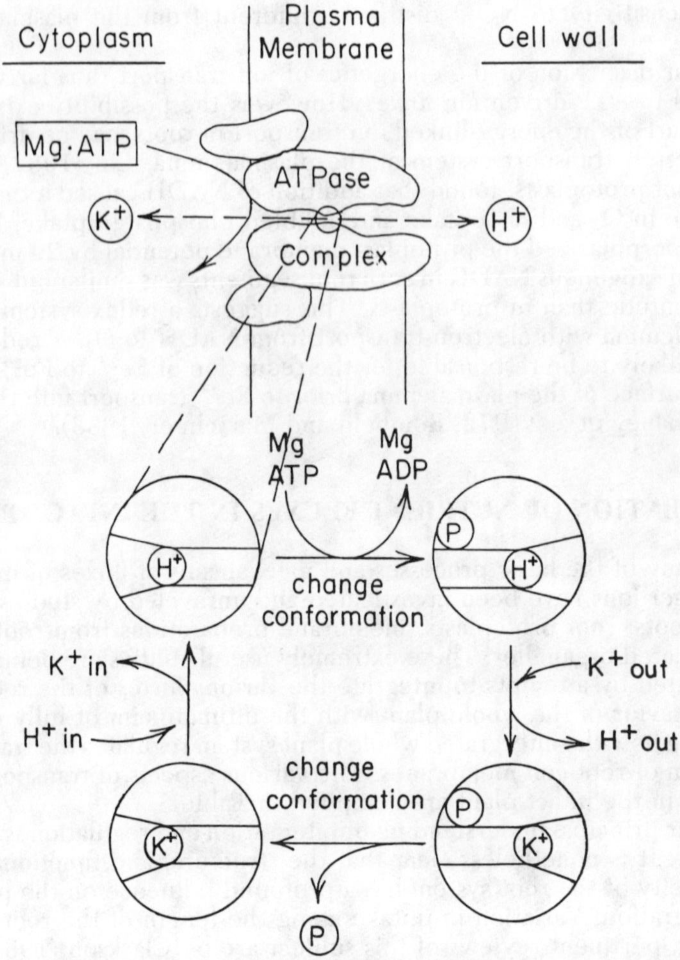

Fig. 6. Possible structure and function of the plasmalemma ATPase in plant cells (Leonard, 1983).

of which was influenced by cations and anions (Leigh and Walker, 1980; Walker and Leigh, 1981). Particularly, this enzyme was stimulated most by Cl^- but inhibited by NO_3^- and, in contrast to the plasmalemma ATPase, it was insensitive to vanadate. Other supporting evidence was derived from ATP-dependent H^+ transport into membrane vesicles that were possibly derived from vacuoles, as indicated by Cl^- stimulation, NO_3^- inhibition, and lack of sensitivity to vanadate (DuPont et al., 1982; Stout and Cleland, 1982; Churchill and Sze, 1983; O'Neill et al., 1983). Solute transport systems at the tonoplast have been more fully reviewed by Leigh (1983). From these results, the conclusion can be drawn that the tonoplast of root cells appears to contain an ATPase involved in H^+ pumping to the vacuole which, interestingly, appears to have properties (e.g., sensitivity to NO_3^-) distinctly different from the plasmalemma ATPase.

Our discussion of the energetics of ion transport thus far was restricted to ATP-driven ion fluxes. However, the possibility exists that some part of the energy-linked ion transport in roots may be driven by an electron transport system at the plasmalemma. Lin (1982), using corn root protoplasts, found that addition of NADH caused a threefold increase in O_2 and K^+ uptake and doubled phosphate uptake; NADH also hyperpolarized the protoplast membrane potential by 20 mV. The effect of exogenous NADH in corn root segments was similar but smaller in magnitude than in protoplasts. This suggests a redox system in the plasmalemma with electron transport from NADH to O_2. A redox system is likely to be responsible for the reduction of Fe^{3+} to Fe^{2+} at the outer surface of the plasmalemma prior to Fe^{2+} transport into the root cell (Chaney et al., 1972; Römheld and Marschner, 1983).

REGULATION OF NUTRIENT FLUXES IN THE INTACT PLANT

Many of the basic processes and mechanisms of fluxes of nutrient and other ions have been investigated and unraveled by studies of excised roots, root protoplasts, membrane preparations from root cells, and isolated organelles. These extremely useful studies are being complemented by attempts to integrate the various fluxes of the root into the behavior of the whole plant with the ultimate aim of fully understanding how the intergrated whole plant system regulates the transport function of root cell membranes. The various aspects of transport regulation in the intact plant are compiled in Table 3.

Our principal understanding of integration and regulation is rather sketchy. It is nonetheless clear that the structural and functional heterogeneity of the root system has a profound influence on the pattern of integration. Variation in uptake along the length of the root is important; pertinent reviews of this subject are by Clarkson (1981) and Lüttge (1983). Also important is the question whether ion uptake occurs primarily at the root periphery or also involves the entire cortex. In thin, fibrous roots like wheat (*Triticum aestivum* L.) and barley which

Table 3. Various components of transport regulation in the intact plant.

a. Root:
   - variation in uptake along the length of the root
   - apoplasmic vs. symplasmic transport
   - compartmentation in cells and tissues
   - transfer cells
   - nutrient status of the root

b. Shoot-root interactions:
   - translocation and circulation in the plant
   - photosynthate supply from the shoot
   - regulation by the shoot of ion transport from root to shoot (relative growth rate of shoot; hormones, "signals")

c. Genetic control

d. Regulatory responses to environmental stresses.

only contain three or four cortical cell layers diffusion through the apoplasm to the endodermis seems to be rapid and ion uptake may involve the whole cortex. In the thicker root of corn (and possibly other species of similar root cortex structure), however, ion uptake may be primarily located at the epidermis. In particular, at relatively low ion concentrations in the medium it appears likely that the cells of the root periphery scavenge most of the ions that move towards the surface of relatively thick roots. Thus, radial ion transport to the stele would be primarily through the symplasm. Recent support of this concept came from studies of the localized effect of sulfhydryl reagents on ion uptake at the periphery of corn roots (Kochian and Lucas, 1983). Earlier, it was demonstrated by X-ray microanalysis that epidermal cells of corn roots accumulate K relative to the cortex (Läuchli, 1967; Läuchli et al., 1971). Other evidence supporting this concept was cited by Kochian and Lucas (1983). Furthermore, the discovery of transfer cell induction in the root epidermis caused by Fe stress (review: Kramer, 1981) adds weight to the importance of the root periphery in ion uptake. Cortical cells, however, may become more important at higher ion concentrations in the substrate.

Regarding transport regulation in the root itself, ion compartmentation and the nutrient status of the root play an outstanding role; these aspects will be briefly reviewed in the next two sections of this chapter. The importance of the shoot in regulation of transport processes in the root will then be assessed in the section Shoot-Root Interactions. However, genetic control of membrane transport and regulatory responses to environmental stresses are beyond the scope of this review. For those features of transport regulation not treated here the reader is referred to pertinent papers in the *Encyclopedia of Plant Physiology*, New Series, Vol. 2A and 2B (Lüttge and Pitman, 1976) and Vol. 15A and 15B (Läuchli and Bieleski, 1983) and in *Annual Review of Plant Physiology* and related series.

Compartmentation in Root Cells

It has long been recognized that the three principal compartments of mature root cells which clearly differ greatly in their ionic composition are the cell wall, cytoplasm, and vacuole. Cells of the epidermis, cortex, endodermis, and stelar parenchyma are organized in these three main compartments. The cytoplasmic compartment is not only the site of major metabolic activity in the cell, it is also part of the symplasmic pathway and thus clearly distinct from the vacuole which serves as a nutrient reservoir, the principal osmotic space in the ionic relations of the cell, and also contains many enzymes (Matile, 1978). Structurally and functionally the cytoplasmic compartment is heterogeneous, the ER (Fig. 1) and the mitochondria being very important subcompartments relative to the ionic relations of the cell. Unfortunately, the techniques presently available for quantitative study of ion compartmentation in root cells do not provide sufficient resolution to analyze separately the cytosol, ER and mitochondria of the cytoplasm without cell fractionation; hence, the cytoplasm is commonly treated as a single compartment.

The techniques for quantitative localization of ions in the cell compartments have been reviewed by Flowers and Läuchli (1983) for K and Na; in principle they are also applicable to other ions. X-ray microanalysis provides the most direct information (see for example Heinrich, 1981 for principles of X-ray microanalysis). Its resolution makes the method amenable to quantitative analyses of cytoplasm and vacuole of root cells (Pitman et al., 1981a). An example is shown in Fig. 7. The data demonstrate that there are distinct differences in the distribution of mineral elements between cytoplasm and vacuole of cortex cells of barley roots. It is obvious that the bulk of Na is sequestered in the vacuole while K is uniformly distributed between cytoplasm and vacuole under the condition of ample K supply. In $CaSO_4$ grown roots (low-salt roots), however, the K concentration was estimated to vary between about 70 and 120 m*M* in the cytoplasm and 9 to 12 m*M* in the vacuole. The preferential localization of K in the cytoplasm of low-salt roots is in line with expectations based on our knowledge of K functions (Läuchli and Pflüger, 1979; Wyn Jones and Pollard, 1983)

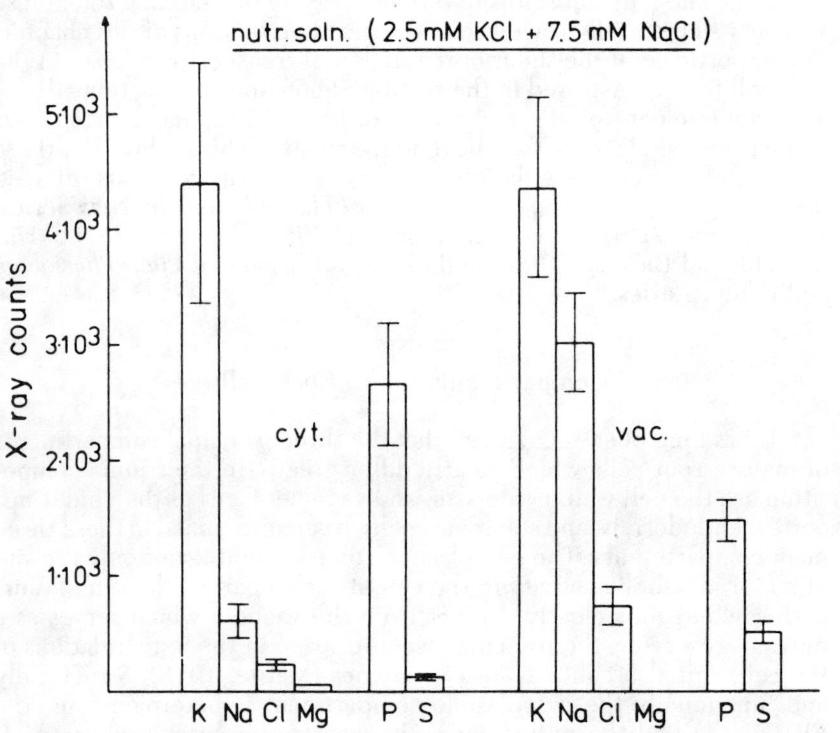

Fig. 7. Compartmentation of six elements in roots of barley seedlings measured by X-ray microanalysis. (X-ray counts are a relative measure of elemental concentrations.) Seedlings grown in complete nutrient solution containing 2.5 m*M* KCl and 7.5 m*M* NaCl (Pitman et al., 1981a).

and compares well with data by Jeschke and Stelter (1976) using flameless atomic absorption spectroscopy. Modern X-ray microanalysis instrumentation with much improved resolution and sensitivity will contribute substantially to our knowledge of quantitative ion compartmentation in root cells and its role in regulation of ion uptake.

Regulation by Nutrient Status of the Root

Ion uptake is profoundly influenced by the nutrient status of the root. Influx of ions appears to be negatively correlated with the tissue concentration of the particular ion (e.g., Cram, 1973; Glass, 1975). Taking tissue and substrate concentration simultaneously into consideration, Siddiqi and Glass (1982) derived an equation which is an extension of the Michaelis-Menten Eq. [2]:

$$v_{(c_i)} = (\text{Max } V_{max}\ e^{-b\,c_i}\ c_o)/(\text{Min } K_m\ e^{b'\,c_i} + c_o) \qquad [9]$$

(modified from Siddiqi and Glass, 1982). $v_{(c_i)}$ is the influx at a given tissue concentration c_i, Max V_{max} is the maximum V_{max} and Min K_m is the minimum K_m when V_{max} and K_m, respectively, are determined at varying tissue concentration c_i, b is the slope relating V_{max} and b′ is the slope relating K_m to the tissue concentration c_i, and c_o is the substrate concentration. Equation [9] was tested for K^+ influx in barley roots (Fig. 8); the observed influx data were found to agree well with the estimated influx data based on Eq. [9]. This quantitative model needs to be tested

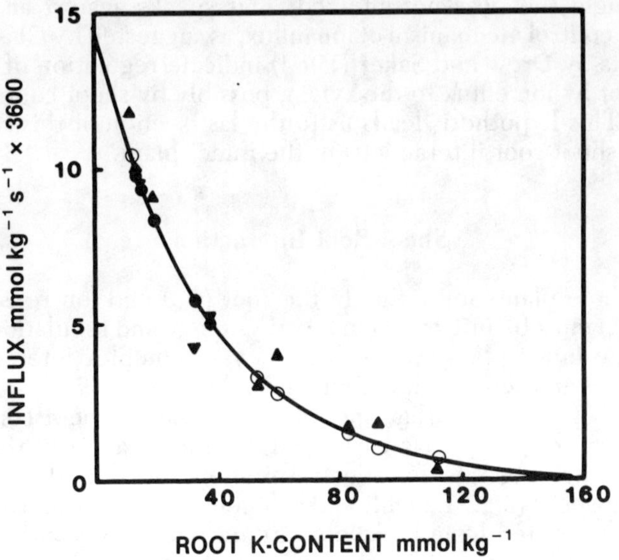

Fig. 8. Dependence of K^+ influx upon K^+ concentration in the root. Closed triangle: measured influx of K^+ (^{86}Rb) from 60 μM K^+; open circle: influx values estimated by Eq. [9]. The line was drawn after exponential curve fitting (r^2= 0.93) (Siddiqi and Glass, 1982).

for other nutrient ions and its relevance be expanded to incorporate effects on influx of a particular ion by varying substrate and tissue concentrations of other ions. Lefebvre and Glass (1982) indeed found that phosphate influx in barley roots declined with increased phosphate concentration of the roots, confirming in principle Glass' earlier results on K^+. Glass (1976, reviews: 1983, 1984) therefore proposed a negative allosteric feedback mechanism in which ion influx in roots is regulated by the internal concentration of the given ion in the root. On the other hand, it is well known that not only the ion in question but other internal ions can have a large effect on the rate of influx of a given ion. Hence, the negative feedback concept may be too narrow.

In a detailed study on changes in the kinetics of K^+ and phosphate influx in nutrient-starved roots of intact barley plants, Drew et al. (1984) demonstrated increased V_{max} in response to K^+ or phosphate deprivation. A possible increase in V_{max} for phosphate transport was already found earlier when only part of the root was exposed to this ion (Drew and Saker, 1978). In addition, K^+ deprivation caused a decrease in K_m, while no such change in K_m was observed for phosphate (Drew et al., 1984). These changes in the kinetics of K^+ and phosphate influx were associated with lower concentrations of these ions in roots and shoots. Using plants with divided root systems where a single seminal root was supplied with all the nutrient ions, while the remaining roots were temporarily subjected to a nutrient deficiency (K^+, Cl^-, phosphate) treatment, ion influx and transport to the shoot by the single root was stimulated by the nutrient deficiency treatment to the remaining roots (Drew and Saker, 1984). However, the concentration of the given ion in the single root was not affected. This speaks against an allosteric feedback control mechanism of ion influx, as suggested by Glass. Rather, the results by Drew and Saker (1984) indicate regulation of ion influx in the root by ion efflux to the xylem, possibly by shoot control via the phloem. This hypothesis leads us to the last section of this review devoted to shoot-root interactions in the intact plant.

Shoot-Root Interactions

In whole plants ion influx in the root (Φ_{oc}) and ion release to the xylem (Φ_{cx}) must be integrated in the physiology and regulation of transport in the intact plant system with its very complex interactions between shoot and root (reviews: Pitman and Cram, 1977; Lüttge, 1983). It is clearly recognized that one important way of shoot control over transport in the root is via carbohydrate translocation in the phloem to supply the substrate for root respiration and thus the energy for active transport (e.g., Läuchli and Pflüger, 1979). Carbohydrates, although transmitted from the shoot, are unspecific substances in terms of control over the individual ion fluxes in the root. Indirectly, this was borne out by the potent inhibition of Φ_{cx} by the uncoupler carbonylcyanide *m*-chlorophenylhydrazone (CCCP) (Läuchli and Epstein, 1971; Pitman, 1972). The CCCP has also been found to affect ion uptake by the root and, moreover, it inhibited water flow through the root by reduction of the hydraulic conductivity L_p (Pitman et al., 1981b). Thus, ion fluxes in the root are commonly affected by metabolic inhibitors and depend upon the supply of energy from respiration. But, the fact that Φ_{cx} is very sensitive to CCCP renewed the interest in the question whether ion release to the xylem is an active transport process (Läuchli, 1972). Early considerations of the electrochemical gradients for ions moving from the root periphery to the xylem vessels suggested that ions are transported passively from the living stelar cells to the xylem (e.g., Bowling, 1973; review: Bowling, 1981). Using a different approach, Davis and Higinbotham (1976) obtained results that are not entirely in agreement with Bowling's hypothesis. These authors calculated ion fluxes into and out of the vessels of corn roots and applied the flux ratio equation (Eq. [5]) to distinguish between active and passive ion fluxes. They concluded that release of K^+ to the xylem was against its electrochemical gradient, that is active, but Cl^- was moving to the xylem with the electrochemical gradient.

Potentially, hormones may be more specific in terms of differentially affecting ion uptake and release to the xylem. As reviewed in some detail by Pitman (1977; 1982a), abscisic acid (ABA), cytokinins, the protein synthesis inhibitor cycloheximide and certain amino acid analogues all inhibit Φ_{cx} in barley roots while having little or no effect on short-term ion uptake by the root. In corn roots, however, ABA had a rapid inhibitory effect on K^+ uptake (Shaner et al., 1975). In contrast to CCCP, ABA was demonstrated to reduce specifically Φ_{cx} in barley without affecting water flow (Pitman and Wellfare, 1978), although in some other plants ABA appeared to increase the hydraulic conductivity of the root (e.g., Collins and Kerrigan, 1973). It is not known whether ABA is the controlling shoot signal transmitted to the root and possibly regulating ion release to the xylem. An alternative view is that ABA indirectly affects transport and that ion release to the xylem may be regulated by cytokinins (cf. Pitman et al., 1974; Pitman and Cram, 1977; Jeschke, 1982).

Experiments involving the following amino acid analogues were particularly revealing:

amino acid	analogue
L-phenylalanine	DL-*p*-fluorophenylalanine (FPA)
L-proline	L-azetidine-2-carboxylic acid (AZ).

(Amino acid analogues may lead to the formation of ineffective proteins during protein synthesis.) The analogue FPA inhibited Φ_{cx} in barley roots without having an effect on ion uptake by the root up to 5 h (Schaefer et al., 1975; Fig. 9). These authors concluded that Φ_{cx} involved a specific protein with a short effective life (based on the time needed to produce flux inhibition by FPA). They also proposed that the site of FPA action is probably at the interface between xylem parenchyma cells and xylem vessels. Further support of this view came from a study by Läuchli et al. (1978) in which it was demonstrated that

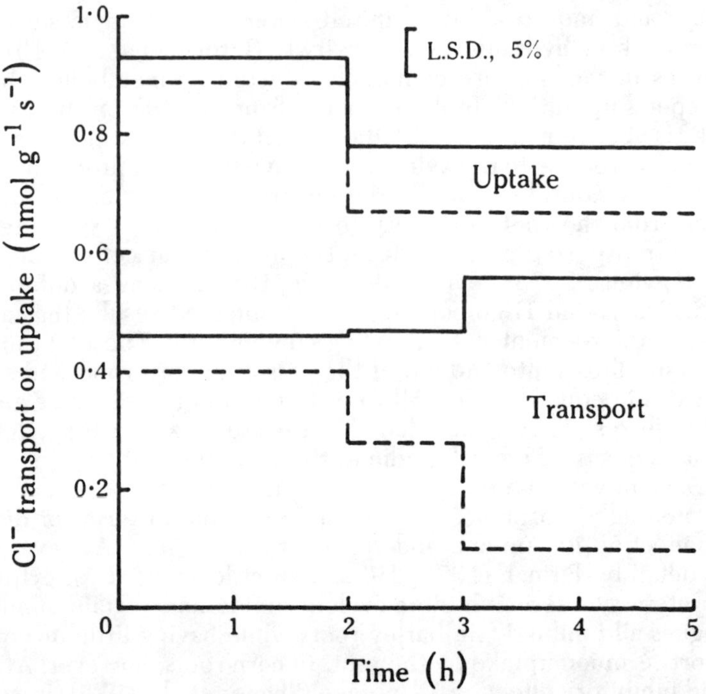

Fig. 9. Effect of the amino acid analogue DL-*p*-fluorophenylaline (FPA) (2 m*M*), applied at t = 0 on rates of $^{36}Cl^-$ uptake and release to the xylem of barley roots. Cl^- concentration in medium: 5 m*M*. Control: solid line; FPA treatment: broken line. (Schaefer et al., 1975).

Φ_{cx} was inhibited by FPA in parts of the root where all the xylem vessels were fully mature (Fig. 10). Root exudation is therefore not due to the activity of developing xylem vessels. The amino acid analogue AZ had an effect similar to FPA (Pitman et al., 1977). Pitman (1980) thereafter showed that the AZ effect was not accompanied by reduction in the hydraulic conductivity L_p of the root (Fig. 11) and AZ was suggested to act on the process of ion release to the xylem. In the latter study L_p was calculated using the equation (Pitman, 1980),

$$L_p = (J_v + J_{v,m})/\pi_m - \pi_o), \qquad [10]$$

where J_v is the rate of fluid exudation from the xylem, $J_{v,m}$ is the rate of reverse flow out of the root to a 300 m*M* mannitol solution, and π_m and π_o are the osmotic pressures of the mannitol and nutrient solutions, respectively. J_v and $J_{v,m}$ are expressed as $mm^3\ m^{-2}\ s^{-1}$, and L_p as $mm^3\ m^{-2}\ s^{-1}\ MPa^{-1}$. The reflection coefficient σ for mannitol was considered to be 1.0 and therefore ignored in Eq. [10]. It is pertinent to note that FPA also inhibited more effectively net salt excretion from the salt glands of a salt-tolerant halophyte, *Limonium vulgare* Miller, than salt uptake in the leaf tissue (Jung et al., 1980); salt excretion from a gland is formally analogous to ion release from the root symplasm to the xylem

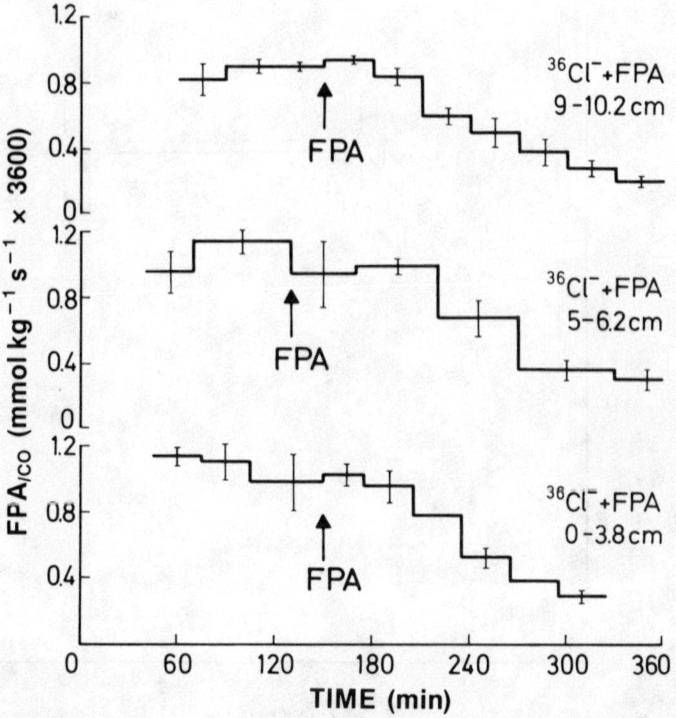

Fig. 10. Effect of FPA (2 m*M*) on rates of $^{36}Cl^-$ release to the xylem of barley roots. Cl^- concentration in medium: 10 m*M*. Tracer added at time = 0; DL-*p*-fluorophenylaline (FPA) added at times indicated by arrows to root regions 0 to 3.8, 5 to 6.2, and 9 to 10.2 cm, respectively. (Metaxylem vessels were fully mature about 1 cm from apex.) Response to FPA given as ratio FPA/control roots (CO), i.e., rates of $^{36}Cl^-$ transport through FPA treated roots over rates through control roots (Läuchli et al., 1978).

vessels. Overall, the studies involving amino acid analogues suggest the involvement of a transport protein, probably located at the xylem parenchyma cells, and likely to be membrane-bound. The involvement of a proteinaceous carrier does not necessarily prove that ion release to the xylem vessels is an active transport process, but it does imply that the plant can regulate this important transport step in response to fluctuations in environmental and internal factors. An attractive possibility is that this protein may be a transport ATPase, as Winter-Sluiter et al. (1977) cytochemically demonstrated K^+-stimulated ATPase activity at plasmalemma and other membranes of xylem parenchyma cells in barley roots. An electrophysiological study of roots of intact white clover (*Trifolium repens* L.) (Dunlop, 1982) yielded results emphasizing the special role of xylem parenchyma cells in regulation of ion release to the xylem. The electrical potential differences (PD) in the xylem parenchyma cells were about 10 mV more negative than in the other living root cells, and the PD in the xylem vessels was about 90 mV less negative than in the adjacent xylem parenchyma cells. Moreover, FPA depolarized PD in the xylem parenchyma by 10 mV and in the xylem

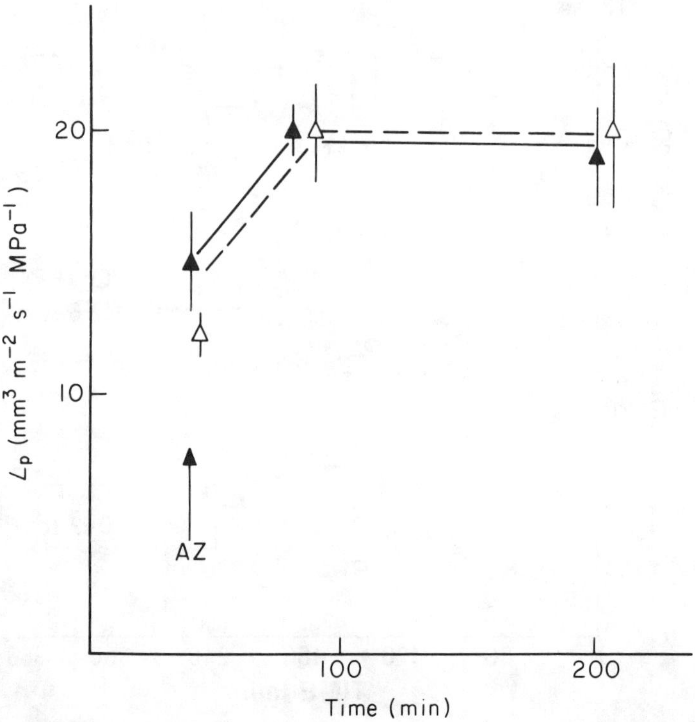

Fig. 11. The effect of L-azetidine-2-carboxylic acid (AZ) (50 *μM*) on the hydraulic conductivity L_p in barley roots. The AZ added at time indicated by arrow. Control: closed symbols; AZ treatment: open symbols (Pitman, 1980).

vessels by 50 mV. It will be exciting to reconcile these recent data on the transport function of xylem parenchyma cells with Hanson's (1978) application of the chemiosmotic concept to ion transport through roots, the idea being that there may be H^+-coupled transport at both the sites of ion uptake at the root periphery and at the site of ion release to the xylem. Clarkson et al. (1984) recently presented evidence for an H^+ pump in the stele of onion (*Allium cepa* L.) roots.

In summary, it appears that the shoot interacts with ion transport in the root in a variety of ways. The results from the hormone and amino acid analogue studies suggest that shoot regulation of ion influx in the root may be different in kind from that of ion release to the xylem. In operational terms, shoot control of transport in the root is likely to be exerted by information transmitted from shoot to root and involving signal(s) of as yet unknown chemical nature (Pitman and Cram, 1977; Pitman, 1982b).

ACKNOWLEDGMENT

I thank Malcolm Drew and Emanuel Epstein for reviewing the manuscript.

REFERENCES

1. Anderson, W.P. 1976. Transport through roots. p. 129–156. *In* U. Lüttge and M.G. Pitman (ed.) Encyclopedia of plant physiology. New Series, Vol. 2, Part B. Springer-Verlag, New York.
2. Bellando, M., A. Trotta, A. Bonetti, R. Colombo, P. Lado, and E. Marré. 1979. Dissociation of H^+ extrusion from K^+ uptake by means of lipophilic cations. Plant Cell Environ. 2:39–47.
3. Ben-Zioni, A., Y. Vaadia, and S.H. Lips. 1971. Nitrate uptake by roots as regulated by nitrate reduction products of the shoot. Physiol. Plant. 24:288–290.
4. Bowling, D.J.F. 1973. The origin of the trans-root potential and the transfer of ions in the xylem of sunflower roots. p. 483–491. *In* W.P. Anderson (ed.) Ion transport in plants. Academic Press, New York.
5. ----. 1981. Release of ions to the xylem in roots. Physiol. Plant. 53:392–397.
6. Briskin, D.P., and R.T. Leonard. 1980. Isolation of tonoplast vesicles from tobacco protoplasts. Plant Physiol. 66:684–687.
7. Chaney, R.L., J.C. Brown, and J.C. Tiffin. 1972. Obligatory reduction of ferric chelates in iron uptake by soybeans. Plant Physiol. 50:208–213.
8. Chastain, C.J., and J.B. Hanson. 1982. Control of proton efflux from corn root tissue by an injury-sensing mechanism. Plant Sci. Lett. 24:97–104.
9. Churchill, K.A., and H. Sze. 1983. Anion-sensitive, H^+-pumping ATPase in membrane vesicles from oat roots. Plant Physiol. 71:610–617.
10. Clarkson, D.T. 1981. Nutrient interception and transport by root systems. p. 307–330. *In* C.B. Johnson (ed.) Physiological processes limiting plant productivity. Butterworths, Massachusetts.
11. ----, and J.B. Hanson. 1980. The mineral nutrition of higher plants. Ann. Rev. Plant Physiol. 31:239–298.
12. ----, L. Williams, and J.B. Hanson. 1984. pH regulation of the xylem sap in onion roots: the role of the stelar proton pump. Plant Physiol. Suppl. 75:44.
13. Collins, J.C., and A.P. Kerrigan. 1973. Hormonal control of ion movements in the plant root. p. 589–593. *In* W.P. Anderson (ed.) Ion transport in plants. Academic Press, New York.
14. Colombo, R., A. Bonetti, R. Cerana, and P. Lado. 1981. Effect of plasmalemma ATPase inhibitors, diethylstilbestrol and orthovanadate, on fusicoccin-induced H^+ extrusion in maize roots. Plant Sci. Lett. 21:305–315.
15. Cram, W.J. 1973. Internal factors regulating nitrate and chloride influx. J. Exp. Bot. 24:328–341.
16. Davis, R.F., and N. Higinbotham. 1976. Electrochemical gradients and K^+ and Cl^- fluxes in excised corn roots. Plant Physiol. 57:129–136.
17. Dieter, P., and D. Marmé. 1980. Ca^{2+} transport in mitochondrial and microsomal fractions from higher plants. Planta 150:1–8.
18. Drew, M.C., and L.R. Saker. 1978. Nutrient supply and the growth of the seminal root system in barley. II. Compensatory increases in growth of lateral roots and in rates of phosphate uptake in response to a localized supply of phosphate. J. Exp. Bot. 29:435–451.
19. ----, and ----. 1984. Uptake and long-distance transport of phosphate, potassium and chloride in relation to internal ion concentrations in barley: evidence of non-allosteric regulation. Planta 160:500–507.
20. ----, ----, S.A. Barber, and W.A. Jenkins. 1984. Changes in the kinetics of phosphate and potassium absorption in nutrient-deficient barley roots measured by a solution depletion technique. Planta 160:490–499.
21. Dunlop, J. 1982. Membrane potentials in the xylem in roots of intact plants. J. Exp. Bot. 33:910–918.
22. ----, and D.J.F. Bowling. 1978. Uptake of phosphate by white clover. II. The effect of pH on the electrogenic phosphate pump. J. Exp. Bot. 29:1147–1153.

23. DuPont, F.M., D.L. Giorgi, and R.M. Spanswick. 1982. Characterization of a proton-translocating ATPase in microsomal vesicles from corn roots. Plant Physiol. 70:1694–1699.
24. Epstein, E. 1972. Mineral nutrition of plants: Principles and perspectives. John Wiley & Sons, New York.
25. ----. 1976. Kinetics of ion transport and the carrier concept. p. 70–94. *In* U. Lüttge and M.G. Pitman (ed.) Encyclopedia of plant physiology. New Series, Vol. 2, Part B. Springer-Verlag, New York.
26. ----, and C.E. Hagen. 1952. A kinetic study of the absorption of alkali cations by barley roots. Plant Physiol. 27:457–474.
27. Fisher, J.D., D. Hansen, and T.K. Hodges. 1970. Correlation between ion fluxes and ion-stimulated adenosine triphosphatase activity of plant roots. Plant Physiol. 46:812–814.
28. Flowers, T.J., and A. Läuchli. 1983. Sodium versus potassium: substitution and compartmentation. p. 651–681. *In* A. Läuchli and R.L. Bieleski (ed.) Encyclopedia of plant physiology. New Series, Vol. 15 B. Springer-Verlag, New York.
29. Gallagher, S.R., and R.T. Leonard. 1982. Effect of vanadate, molybdate, and azide on membrane-associated ATPase and soluble phosphatase activities of corn roots. Plant Physiol. 70:1335–1340.
30. Glass, A. 1975. The regulation of potassium absorption in barley roots. Plant Physiol. 56:377–380.
31. ----. 1976. Regulation of potassium absorption in barley roots: an allosteric model. Plant Physiol. 58:33–37.
32. ----. 1983. Regulation of ion transport. Ann. Rev. Plant Physiol. 34:311–326.
33. ----, and M.Y. Siddiqi. 1982. Cation-stimulated H^+ efflux by intact roots of barley. Plant Cell Environ. 5:385–393.
34. ----, and ----. 1984. The control of nutrient uptake rates in relation to the inorganic composition of plants. p. 103–147. *In* P.B. Tinker and A. Läuschli (ed.) Advances in plant nutrition. Vol. 1. Praeger Publishers, New York.
35. ----, ----, and K.I. Giles. 1981. Correlations between potassium uptake and hydrogen efflux in barley varieties. A potential screening method for the isolation of nutrient efficient lines. Plant Physiol. 68:457–459.
36. Hanson, J.B. 1978. Application of the chemiosmotic hypothesis to ion transport across the root. Plant Physiol. 62:402–405.
37. ----. 1984. The functions of calcium in plant nutrition. p. 149–208. *In* P.B. Tinker and A. Läuchli (ed.) Advances in plant nutrition. Vol. 1. Praeger Publishers, New York.
38. Heinrich, K.F.J. 1981. Electron beam X-ray microanalysis. Van Nostrand Reinhold Co., New York.
39. Hodges, T.K. 1973. Ion absorption by plant roots. Adv. Agron. 25:163–207.
40. ----. 1976. ATPases associated with membranes of plant cells. p. 260–283. *In* U. Lüttge and M.G. Pitman (ed.) Encyclopedia of plant physiology. New Series, Vol. 2, Part A. Springer-Verlag, New York.
41. Jackson, P.C. 1982. Differences between effects of undissociated and anionic 2, 4 - dinitrophenol on permeability of barley roots. Plant Physiol. 70:1373–1379.
42. Jeschke, W.D. 1980a. Involvement of proton fluxes in the K^+-Na^+ selectivity at the plasmalemma; K^+-dependent net extrusion of sodium in barley roots and the effect of anions and pH on sodium fluxes. Z. Pflanzenphysiol. 98:155–175.
43. ----. 1980b. Roots: cation selectivity and compartmentation, involvement of protons and regulation. p. 17–28. *In* R.M. Spanswick et al. (ed.) Plant membrane transport: Current conceptual issues. Elsevier/North-Holland Biomedical Press, Amsterdam.
44. Jeschke, W.D. 1982. Shoot-dependent regulation of sodium and potassium fluxes in roots of whole barley seedlings. J. Exp. Bot. 33:601–618.
45. ----, and W. Stelter. 1976. Measurement of longitudinal ion profiles in single roots of *Hordeum* and *Atriplex* by use of flameless atomic absorption spectroscopy. Planta 128:107–112.

46. Jung, K.-D., E. Ball, and U. Lüttge. 1980. Inhibition of sugar and salt elimination by glands with the amino acid analog *p*-fluorophenylalanine. Ann. Bot. 45:351–356.

47. Kochian, L.V., and W.J. Lucas. 1982. Potassium transport in corn roots. I. Resolution of kinetics into a saturable and linear component. Plant Physiol. 70:1723–1731.

48. ----, and ----. 1983. Potassium transport in corn roots. II. The significance of the root periphery. Plant Physiol. 73:208–215.

49. Kramer, D. 1981. Structure and function in absorption and transport of nutrients. p. 303–307. *In* R. Brouwer et al. (ed.) Structure and function of plant roots. Martinus Nijhoff/Dr. W. Junk Publ., The Hague.

50. Läuchli, A. 1967. Untersuchungenüber verteilung und transport von ionen in pflanzengeweben mit der röntgen-mikrosonde. I. Versuche an vegetativen organen von *Zea mays*. Planta 75:185–206.

51. ----. 1972. Translocation of inorganic solutes. Ann. Rev. Plant Physiol. 23:197–218.

52. ----. 1976a. Apoplasmic transport in tissues. p. 3–34. *In* U. Lüttge and M.G. Pitman (ed.) Encyclopedia of plant physiology. New Series, Vol. 2, Part B. Springer-Verlag, New York.

53. ----. 1976b. Symplasmic transport and ion release to the xylem. p. 101–112. *In* I.F. Wardlaw and J.B. Passioura (ed.) Transport and transfer processes in plants. Academic Press, New York.

54. ----, and R.L. Bieleski (ed.) 1983. Inorganic plant nutrition. Encyclopedia of plant physiology. New Series, Vol. 15 A and B. Springer-Verlag, New York.

55. ----, and E. Epstein. 1971. Lateral transport of ions into the xylem of corn roots. I. Kinetics and energetics. Plant Physiol. 48:111–117.

56. ----, and R. Pflüger. 1979. Potassium transport through plant cell membranes and metabolic role of potassium in plants. p. 111–163. *In* Potassium research—review and trends. Int. Potash Inst., Bern/Switzerland.

57. ----, M.G. Pitman, U. Lüttge, D. Kramer, and E. Ball. 1978. Are developing xylem vessels the site of ion exudation from root to shoot? Plant Cell Environ. 1:217–223.

58. ----, A.R. Spurr, and E. Epstein. 1971. Lateral transport of ions into the xylem of corn roots. II. Evaluation of a stelar pump. Plant Physiol. 48:118–124.

59. Lefebvre, D.D., and A.D.M. Glass. 1982. Regulation of phosphate influx in barley roots: effects of phosphate deprivation and reduction of influx with provision of orthophosphate. Physiol. Plant 54:199–206.

60. Leigh, R.A. 1983. Methods, progress and potential for the use of isolated vacuoles in studies of solute transport in higher plant cells. Physiol. Plant. 57:390–396.

61. ----, and R.R. Walker. 1980. ATPase and acid phosphatase activities associated with vacuoles isolated from storage roots of red beet (*Beta vulgaris* L.). Planta 150:222–229.

62. Leonard, R.T. 1982. The plasma membrane ATPase of plant cells: cation or proton pump? p. 633–637. *In* A.N. Martonosi (ed.) Membranes and transport. Vol. 2. Plenum Press, New York.

63. ----. 1983. Potassium transport and the plasma membrane-ATPase in plants. p. 71–86. *In* D.A. Robb and W.S. Pierpoint (ed.) Metals and micronutrients: Uptake and utilization by plants. Academic Press, New York.

64. ----. 1984. Membrane-associated ATPases and nutrient absorption by roots. p. 209–240. *In* P.B. Tinker and A. Läuchli (ed.) Advances in plant nutrition. Vol. 1. Praeger Publishers, New York.

65. ----, and T.K. Hodges. 1973. Characterization of plasma membrane-associated adenosine triphosphatase activity of oat roots. Plant Physiol. 52:6–12.

66. ----, and C.W. Hotchkiss. 1976. Cation-stimulated adenosine triphosphatase activity and cation transport in corn roots. Plant Physiol. 58:331–335.

67. Lin, W. 1982. Response of corn root protoplasts to exogenous reduced nicotineamide adenine dinucleotide: Oxygen consumption, ion uptake, and membrane potential. Proc. Natl. Acad. Sci. USA 70:3773–3776.
68. Lüttge, U. 1983. Import and export of mineral nutrients in plant roots. p. 181–211. *In* A. Läuchli and R.L. Bieleski (ed.) Encyclopedia of plant physiology. New Series, Vol. 15A. Springer-Verlag, New York.
69. ----, and N. Higinbotham. 1979. Transport in plants. Springer-Verlag, New York.
70. ----, and M.G. Pitman. 1976. Transport in plants. II. Encyclopedia of plant physiology. New Series, Vol. 2A and 2B. Springer-Verlag, New York.
71. Macklon, A.E.S. 1975. Cortical cell fluxes and transport to the stele in excised root segments of *Allium cepa* L. II. Calcium. Planta 122:131–141.
72. ----, and A. Sim. 1981. Cortical cell fluxes and transport to the stele in excised root segments of *Allium cepa* L. IV. Calcium as affected by its external concentration. Planta 152:381–387.
73. Marmé, D. 1983. Calcium transport and function. p. 599–625. *In* A. Läuchli and R.L. Bieleski (ed.) Encyclopedia of plant physiology. New Series, Vol. 15 B. Springer-Verlag, New York.
74. Marré, E. 1979. Fusicoccin: A tool in plant physiology. Ann. Rev. Plant Physiol. 30:273–288.
75. Matile, P. 1978. Biochemistry and function of vacuoles. Ann. Rev. Plant Physiol. 29:193–213.
76. Nissen, P., and Ø. Nissen. 1983. Validity of the multiphasic concept of ion absorption in plants. Physiol. Plant 57:47–56.
77. O'Neill, S.D., A.B. Bennett, and R.M. Spanswick. 1983. Characterization of a NO_3^--sensitive H^+-ATPase from corn roots. Plant Physiol. 72:837–846.
78. Pitman, M.G. 1969. Simulation of Cl^- uptake by low-salt barley roots as a test of models of salt uptake. Plant Physiol. 44:1417–1427.
79. ----. 1972. Uptake and transport of ions in barley seedlings. II. Evidence for two active stages in transport to the shoot. Aust. J. Biol. Sci. 25:243–257.
80. ----. 1976. Ion uptake by plant roots. p. 95–128. *In* U. Lüttge and M.G. Pitman (ed.) Encyclopedia of plant physiology. New Series, Vol. 2, Part B. Springer-Verlag, New York.
81. ----. 1977. Ion transport into the xylem. Ann. Rev. Plant Physiol. 28:71–88.
82. ----. 1980. Measurement of hydraulic conductivity of barley roots during inhibition of ion transport by azetidine-2-carboxylic acid. Plant Cell Environ. 3:59–61.
83. ----. 1982a. Transport across plant roots. Q. Rev. Biophys. 15:481–554.
84. ----. 1982b. Ion transport across the root. p. 657–662. *In* A.N. Martonosi (ed.) Membranes and transport. Vol. 2. Plenum Press, New York.
85. ----, and W.J. Cram. 1977. Regulation of ion content in whole plants. p. 391–424. *In* D.H. Jennings (ed.) Integration of activity in the higher plant. Symp. Soc. Exp. Biol. no. 31. Cambridge University Press, New York.
86. ----, A. Läuchli, and R. Stelzer. 1981a. Ion distribution in roots of barley seedlings measured by electron probe X-ray microanalysis. Plant Physiol. 68:673–679.
87. ----, U. Lüttge, A. Läuchli, and E. Ball. 1974. Action of abscisic acid on ion transport as affected by root temperature and nutrient status. J. Exp. Bot. 25:147–155.
88. ----, N. Schaefer, and R.A. Wildes. 1975. Stimulation of H^+ efflux and cation uptake by fusicoccin in barley roots. Plant Sci. Lett. 4:323–329.
89. ----, and D. Wellfare. 1978. Inhibition of ion transport in excised barley roots by abscisic acid; relation to water permeability of the roots. J. Exp. Bot. 29:1125–1138.
90. ----, ----, and C. Carter. 1981b. Reduction of hydraulic conductivity during inhibition of exudation from excised maize and barley roots. Plant Physiol. 67:802–808.
91. ----, R.A. Wildes, N. Schaefer, and D. Wellfare. 1977. Effect of azetidine-2 carboxylic acid on ion uptake and ion release to the xylem of excised barley roots. Plant Physiol. 60:240–246.

92. Poole, R.J. 1978. Energy coupling for membrane transport. Ann. Rev. Plant Physiol. 29:437–460.

93. Römheld, V., and H. Marschner. 1983. Mechanism of iron uptake by peanut plants. I. Fe^{III} reduction, chelate splitting, and release of phenolics. Plant Physiol. 71:949–954.

94. Schaefer, N., R.A. Wildes, and M.G. Pitman. 1975. Inhibition by *p*-fluorophenylalanine of protein synthesis and of ion transport across the roots in barley seedlings. Aust. J. Plant Physiol. 2:61–73.

95. Shaner, D.L., S.M. Mertz, and C.J. Arntzen. 1975. Inhibition of ion accumulation in maize roots by abscisic acid. Planta 122:79–90.

96. Siddiqi, M.Y., and A.D.M. Glass. 1982. Simultaneous consideration of tissue and substrate potassium concentration in K^+ uptake kinetics: a model. Plant Physiol. 69:283–285.

97. Spanswick, R.M. 1976. Symplasmic transport in tissues. p. 35–53. *In* U. Lüttge and M.G. Pitman (ed.) Encyclopedia of plant physiology. New Series, Vol. 2, Part B. Springer-Verlag, New York.

98. ----. 1981. Electrogenic ion pumps. Ann. Rev. Plant Physiol. 32:267–289.

99. Stout, R.G., and R.E. Cleland. 1982. Evidence for a Cl^--stimulated Mg ATPase proton pump in oat root membranes. Plant Physiol. 69:798–803.

100. Sze, H. 1980. Nigericin-stimulated ATPase activity in microsomal vesicles of tobacco callus. Proc. Natl. Acad. Sci. USA 77:5904–5908.

101. ----, and K.A. Churchill. 1981. Mg^{2+}/KCl-ATPase of plant plasma membranes is an electrogenic pump. Proc. Natl. Acad. Sci. USA 78:5578–5582.

102. ----, and T.K. Hodges. 1977. Selectivity of alkali cation influx across the plasma membrane of oat roots. Cation specificity of the plasma membrane ATPase. Plant Physiol. 59:641–646.

103. Thibaud, J.B., and C. Grignon. 1981. Mechanism of nitrate uptake in corn roots. Plant Sci. Lett. 22:279–289.

104. Walker, R.R., and R.A. Leigh. 1981. Characterization of a salt-stimulated ATPase activity associated with vacuoles isolated from storage roots of red beet (*Beta vulgaris* L.). Planta 153:140–149.

105. Winter-Sluiter, E., A. Läuchli, and D. Kramer. 1977. Cytochemical localization of K^+-stimulated adenosine triphosphatase activity in xylem parenchyma cells of barley roots. Plant Physiol. 60:923–927.

106. Wyn Jones, R.G., and A. Pollard. 1983. Proteins, enzymes and inorganic ions. p. 528–562. *In* A. Läuchli and R.L. Bieleski (ed.) Encyclopedia of plant physiology. New Series, Vol. 15 B. Springer-Verlag, New York.

107. Zocchi, G., and J.B. Hanson. 1982. Calcium influx into corn roots as a result of cold shock. Plant Physiol. 70:318–319.

Chapter 2

Carbon Utilization and Losses by Plant Root Systems[1]

A. J. M. SMUCKER[2]

Plant root systems have been reported to achieve lengths of up to 71 000 m/plant when growth was uninhibited by environmental stresses or plant competition (Pavlychenko, 1937). Recent reports, however, have indicated that there may be an inverse relationship between total root length and specific root water uptake (Taylor and Klepper, 1978). There are also conflicting reports which relate root length density, axial resistance, and radial resistance to the absorption of ions and water by different cultivars. Traditionally, crop production has been thought to be proportional to the size of the root system. This premise, however, essentially ignores the tremendous investment of photoassimilate energy by plants for the production and maintenance of extensive root systems. Perhaps one reason for accepting this premise is that our knowledge of the morphology and physiology of root systems is inadequate.

When the primary objective of multidisciplinary plant research is to modify plants so that they maximize the utilization of solar energy, soil fertility, and water, then the root systems of both current and future cultivars must be evaluated for their efficient utilization of the limited

---

[1]Contribution from the Michigan Agric. Exp. Stn., East Lansing, MI 48824. Agric. Exp. Stn. J. no. 11084.

[2]Professor of soil biophysics, Dep. of Crop and Soil Sciences, Michigan State University, East Lansing, MI 48824.

Copyright © 1984 Soil Science Society of America, Crop Science Society of America, and American Society of Agronomy, 677 South Segoe Road, Madison, WI 53711. *Roots, Nutrient and Water Influx, and Plant Growth.*

photoassimilates available to each plant. The goal of increasing crop yields by developing superior cultivars can only be reached when both the shoot and root portions of the crop are modified. The extensive and complex root and soil interface is one of the most ignored frontiers of the agronomic sciences and a better understanding of this interface will promote significant improvements in the disciplines of plant genetics, soil amelioration, and plant and soil modeling.

Recently, the root systems of plants have come to be appreciated more by many researchers investigating the genetic modification of plants, the physiology of plants subjected to stress, and the membrane and hormone responses of roots to their environment. With the development of improved methods of analyses and a renewed interest, perhaps we are at the threshold of an enlightenment period where the morphology and physiology of root function is recognized to be more than absorptive surfaces which anchor plants in the soil.

This chapter emphasizes the partitioning and utilization of photoassimilates by plant root systems. Stress-induced modifications of partitioning and some current methods for measuring plant root systems are also presented. Comprehensive reviews of additional plant root functions may be found in Carson (1974), Torrey and Clarkson (1975), Russell (1977), Harley and Russell (1979), Brouwer et al. (1981), and Arkin and Taylor (1981).

DISTRIBUTION OF PHOTOASSIMILATES TO ROOT SYSTEMS

Although the translocation and utilization of photoassimilates by whole plant systems is not clearly defined, they appear to be controlled by the assimilate demands of distant sinks during ontogenetic development or in response to environmental signals to single or multiple biochemical mechanisms operating in the source leaves (Geiger and Giaquinta, 1982). The translocation of C to specific organs is essential for providing metabolic energy and assisting in the communication between plant parts. Although extensive reviews of C budgets have been presented for various plant fractions of soybean (*Glycine max* L.) (Chatterton and Silvius, 1979), *Lupinus* sp. (Pate et al., 1979), and other crops (Hesketh et al., 1982) most emphasize the aboveground portion of the plant.

Biological yield ultimately depends on whether the products of photosynthesis are used to increase photosynthetic capacity, stored, or lost from the plant system by maintenance respiration (McCree, 1982) and exudation by the root system. The crucial process of photoassimilate allocation appears to be a function of the competitive interaction among plant sinks brought about by the environment. It is believed that one of the main reasons for the lack of a positive correlation between CO_2 exchange/unit area of leaves and crop yield, is that a large proportion of photoassimilates are irretrievably lost by the extensive root systems of most plants. The expansive and leaky root systems of contemporary cultivars may be an evolutionary trait carried along by recently developed cultivars (Boyer et al., 1980; Boyer, 1982).

Quantities of Compounds Released by Roots. For many years the quantity of organic compounds released from roots were underestimated primarily because the results were based upon measurements of plant roots grown in nutrient solutions and/or axenic conditions. Studies by Barber and Gunn (1974) showed that mechanical stresses caused roots of cereals to release from 5 to 9% of the plant sugars and amino acids. Subsequent studies by Barber and Martin (1976), Martin (1977), Sauerbeck and Johnen (1977), and Shadan (1980) reported that gaseous and soluble C released by root systems growing in soils or solution culture amounted to 20 to 39% of the C translocated to the roots. These high rates of C lost to the rhizosphere agree with other $^{14}CO_2$ studies where 34 to 45% of the labeled compounds exported to legume roots were reported lost by respiration and exudation (Carr and Pate, 1967; Pate et al., 1979).

Carbon transfer from the root to the rhizosphere results in a net loss of photosynthates from plant root systems. Short- and long-term labeling of plants with ratios of C isotopes have made it possible to separate root and microbial activity in the rhizosphere. Warembourg and Billes (1979) showed a diphasic succession of $^{14}CO_2$ evolution from non-axenic roots of labeled wheat (*Triticum aestivum* L.) plants. The labeling of plants induced a liberation of ^{14}C-substances in the solution with the maximum accumulation occurring after 3 to 6 h. Roots of 14-day-old dry bean (*Phaseolus vulgaris* L.) plants lost up to 21% of the soluble ^{14}C-label translocated from the source leaf during a 2-h pulse and 1-h chase period (Table 1). Although these studies demonstrated a genotype × exudation interaction, it appears that from 26 to 33% of the label transported to the root is lost to the surrounding media primarily as $^{14}CO_2$. It was also reported that $^{14}CO_2$ accounted for less than 5% of the total CO_2 respired by the root system during the 3-h period of measurement.

Further distribution of ^{14}C-label within the roots of dry beans indicated that 29% remained in the mature roots and 11% was utilized for nonrespiratory growth by the actively growing components of roots (Schumacher and Smucker, 1984a). These photoassimilates transported to the roots appear to be used mainly for respiration, modification of cell walls, cell division and extension, and the synthesis of primary metabolites which produce large amounts of metabolic energy.

Table 1. Accumulated biomass and distribution of ^{14}C-label of 17-day-old dry bean cv. Seafarer pulse labeled with ^{14}C-sucrose for a 3-h period. Plants were grown in an aerated solution culture in the greenhouse with light intensities ranging from 200 to 1900 $\mu mol\ m^{-2}\ s^{-1}$. Each value is the mean of four replications (Schumacher and Smucker, 1984a).

	Dry weight	^{14}C-label
	mg plant^{-1}	%
Shoot	495	39
Root	166	40
Solution	--	21

The loss of root exudates appears to be proportional to the concentration of these organic compounds at the root surface since removal of the compounds increases exudation. Prikryl and Vancura (1980) reported that frequent replacements of the nutrient solution of wheat roots increased the exudation of axenic cultures from 20 to 34% during an 8-day period. Additions of the bacteria *Pseudomonas putida* to the formerly axenic culture increased the loss of C from 26 to 52% during the same treatment period (Table 2). Organic C compounds lost by the wheat seedlings of these experiments were directly correlated to root length or dry weight. Their data indicate that leakage by young roots may have resulted from the simple diffusion of substances away from the apoplastic free space of roots. This mechanism for C loss is supported by others who reported increased losses from roots when the permeability of cell membranes was distributed by environmental stresses (Rovira, 1969; Smucker and Erickson, 1976). Johnen and Sauerbeck (1977) reported that from emergence to tillering, approximately 40% of the C fixed by wheat plants became available to soil microorganisms. At later stages, the amount of photosynthetically fixed C accessible to soil microorganisms exceeded the harvested root mass by 2.5-fold. These data indicate that much of the C lost by wheat plants after tillering resulted in the senescence and decomposition of large segments of the root system and may be an indication of the quantities of C lost through the decomposition of roots during the growing season. Minchin and Pate (1973) also reported that when shoots of 21-day-old peas (*Pisum sativum* L.) were labeled with $^{14}CO_2$, 26% of the ^{14}C was incorporated directly into shoot dry matter and 74% was translocated to the root. Nitrogen fixation and nodule growth consumed 20% of the ^{14}CO-photoassimilates. Root growth required 7%. The remaining 47% of the fixed C was lost to the soil.

RESPIRATORY LOSSES OF CARBON BY ROOT SYSTEMS

Photoassimilate requirements for the development of roots are essentially unknown for most plants, yet it appears that they use less than that invested per unit of leaf since roots have fewer specialized tissue which form the complex anatomical structure of leaves. Maintenance respiration costs, however, appear to be higher for roots due to the long distance transport of photoassimilates (Moldau and Karolin, 1977),

Table 2. Relationship between C incorporated into wheat roots and released into sterile and inoculated solution culture media. (Taken from Fig. 7 of Prikryl and Vancura, 1980).†

Time	Carbon exuded: Sterile	Carbon exuded: Inoculated
days	% of total C	
2	20	26
4	34	39
8	34	52

† mg C in solution (mg C in solution and root)$^{-1}$.

vulnerability of relatively unspecialized cells or tissue which interface with the environment resulting in large turnover rates, and the presence of an active cyanide resistant respiration pathway (Lambers et al., 1983). Current knowledge of the priority of photoassimilate translocation indicates that multiple competitive sinks generally control the net growth and metabolism of various plant tissues (Geiger and Giaquinta, 1982). Although the biochemical basis for controlling assimilate partitioning is not completely understood, plant adaptation is highly correlated to the allocation and accumulation of C by various plant parts. How changes in photoassimilate demands by distant root sinks are communicated, during their development or in response to the environment, to the various photochemical mechanisms of leaves requires additional study.

The energy costs of developing and maintaining a functional root system subjected to adverse environments are much greater than in nonstress environments. Stress environments cause more C to be expended for the development of additional structural tissue, maintenance of membrane integrity, production of toxic anaerobic metabolites, exudation losses and the reallocation of C to other portions of the plant (Hale et al., 1971; Smucker and Erickson, 1976; Schumacher and Smucker, 1984a). It is currently unknown how much of this loss can be alleviated through genetic manipulation. Although there is some diversity among isolines, this difference is much less than the increased C losses resulting from environmental stress.

Although the energy and catabolic intermediates produced by respiration are essential to basic functions of the plant root system, it is generally accepted that root biomass would be increased if respiration could be minimized (de Wit et al., 1978; McCree, 1982). The rate-limiting components of root respiration appear to be a function of carbohydrate catabolism from reserves (dry weight) and the current rate of photosynthesis (Hansen and Jenson, 1977). Carbon losses by root respiration have been reported to be 25% of the current photosynthetic rate and 1.5% of the dry weight for clover (*Trifolium* sp.) (McCree, 1970) and up to 13.5% of the root weight of ryegrass (*Lolium multiflorum* L.) (Hansen and Jenson, 1977). Carbon losses by the respiration of nodulated roots appear to approach 50% (Minchin and Pate, 1973) for peas and have been reported to require as much as 11.1 of C per N_2 fixed or over 4 moles of glucose for each mole of N_2 fixed (Dixon et al., 1981). If one assumes the N content of the plant to be approximately 3.5%, then a plant having the revised relative efficiency of 0.75 (Dixon et al., 1981) would respire 70 g of carbohydrate for each growth increment of 100 g of root dry matter. As the relative efficiency is reduced (e.g., during periods of environmental stress) to 0.5 the carbohydrate requirement could exceed 100% of each succeeding growth increment until the stress was removed. The savings to be gained by improving the C efficiency of nodulated legume roots responding to stress environments appear to be tremendous and should be investigated to determine whether reductions in the allocation of C to nodulated roots could be accomplished with little detriment to the integrated plant system. For example, roots of plants which develop in a compensatory manner, by accelerated growth in those areas of the soil profile which exhibit fewer stress conditions, generally cause a reduction in top growth.

Compensatory root growth reflects a greater allocation of C to a portion of the root system. Although this response mechanism is thought to be an advantage for root systems, it results in a greater expenditure of C to accomplish the same task had there been no need for a compensatory response. Recent results from our laboratories indicated that ion absorption, respiration, and water uptake are also greater in the aerated portion of compensatory roots than in the aerated controls (Schumacher and Smucker, 1984b). These combined increases, however, resulted in the less efficient absorption of ions. Oxygen requirements were an average of 23% greater for the absorption of cations by the compensatory roots of three dry edible bean cultivars (Table 3). Oxygen requirements for anion absorption were 7% greater than the control roots. Therefore, since compensatory morphological responses provide the geneticist with a characteristic for selecting plants with superior root systems, specific morphological differences among genotypes should not be made which favor only a portion of the plant to the ultimate detriment of other components of the plant.

Kuiper (1983) reported that the genetic variation and phenotypic plasticity of individual root responses to nutritional levels appear to be an ecological strategy for plant survival. In addition, his data showed a direct relationship between root respiration activities and growth characteristics (e.g., number of seeds) of the plant. Yet, the plasticity of root respiration and these growth characteristics differed significantly among four inbred lines of grass (*Plantago major* L.) and with plant age. He concluded that genotypes producing a low number of seeds represent species with little variation of the physiological characteristics (e.g., growth functions, ion stimulated adenosine triphosphatase (ATPase) activities, root respiration, etc.) and that genotypes producing a high number of seeds were associated with greater plasticity of these physiological characteristics. However, he did not measure the responses of photoassimilate translocation to ionic strengths of the nutrient solution. Therefore, we are unable to conclude whether the plasticity of root respiratory responses to an environmental change

Table 3. Oxygen requirements for cation and anion absorption by compensatory roots of three genotypes of *Phaseolus vulgaris* L. grown in an aerated nutrient solution culture for 14 days using the split-root technique (Schumacher and Smucker, 1984b).

Genotype	Control roots		Compensatory roots	
	Cations†	Anions‡	Cations	Anions
	μM O_2 $(\mu M)^{-1}$			
Seafarer	1.30	1.36	1.45	1.48
Pinto III	1.05	1.11	1.37	1.22
Line 31908	1.10	1.03	1.42	1.03

† Cations measured were Ca^{++}, K^{+}, Mg^{++}, and H^{+}.

‡ Anions measured were NO_3^- and $H_2PO_4^-$.

was due to alternative metabolic pathways of stressed roots or the limitations of available photoassimilate energy to the root system.

The direct relationship between biological yield and the efficient utilization of the limited amount of photoassimilates allocated to the harvested portion of the plant indicates that the rate of photoassimilate flow in plants appears to be controlled by the combined influences of source, path, and sink, while the partitioning of these compounds may be a function of assimilate potentials (Lang and Thorpe, 1983; Thorpe et al., 1983). Such models predict the rate of C flow to be directly related to the number and activity of metabolic sinks which are attenuated by changes in the environment. Results from our laboratories indicate that root systems with fewer actively respiring apical meristems appear to utilize their assimilates more efficiently, especially during periods without environmental stress. However, plant tolerance to soil stresses appears to be related to the rapid allocation and transport of soluble C to the meristems of roots with a concomitant increase in the branching and expansion of the root system (Schumacher and Smucker, 1984a, 1984b).

Recent reports indicate that the net respiration of plant roots is influenced by the cyanide (CN)-resistant alternative pathway which predominates when the cytochrome oxidase pathway is either substrate saturated or adenylate restricted (Lambers et al., 1983). Since the CN-resistant alternative path represents a seemingly unregulated by-pass of two-thirds of the conventional and regulated energy-conserving cytochrome-mediated electron path (Laties, 1982), the mechanisms which engage this energetically inefficient alternate pathway in roots are of substantial physiological significance. Essentially little is known about the C investment into root systems and the incipient respiratory responses to specific biological, chemical, or physical environments. This area of study merits greater emphasis as we improve our technology for investigating the complex rhizosphere.

Root turnover or the death and decomposition of fine roots, may also account for a large loss of plant photoassimilates. Coleman (1976) demonstrated that root production is often the largest C input to the ecosystem as 54% of grass root tips appear to survive for less than 1 month, and may require greater investments of photoassimilates than shoot production even though the remaining quantity of roots at any given time are less than or equal to those of the shoots which have accumulated for the life of the plant.

A schematic of C flow through the excessively large root systems of commercial crops is summarized in Fig. 1. Some C is obviously necessary for the growth, maintenance, and absorption processes of extensive root systems. However, a substantial but larger portion appears to be lost by the large turnover rate of roots (St. Jon and Coleman, 1983), uncontrolled CO_2 respiration and perhaps leakage of HCO_3^- ions (unpublished data). The longer C compounds are retained, either in the form of metabolic recycling or as structural tissue, the greater the efficiency of C use. Therefore, if superior root systems are to be developed for high maintenance agriculture, it may be necessary to select

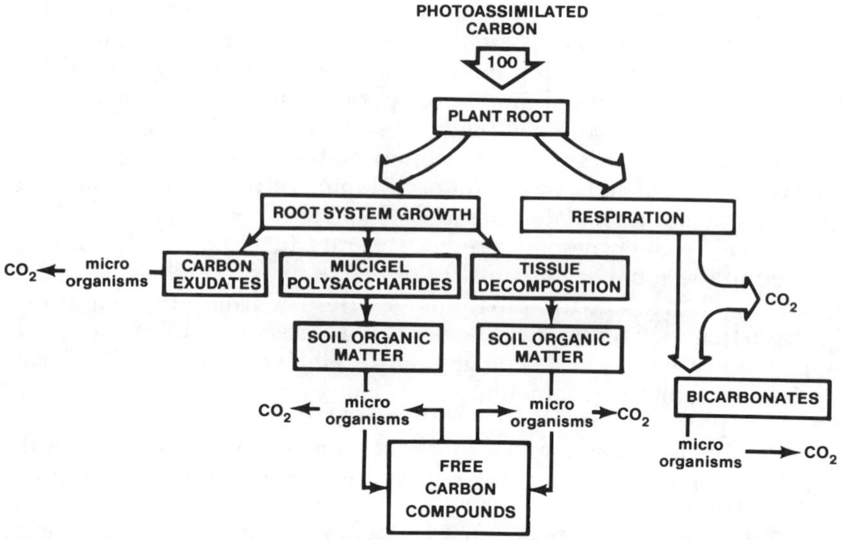

Fig. 1. Flow chart of C utilization by plant roots and the rhizosphere.

roots with very low C turnover rates (e.g., minimum losses of soluble exudates, minimum CN-resistant respiration, minimum number of actively respiring apical meristems, greater resistance to soil decomposing factors and readsorption of CO_2 or bicarbonate ions, etc.).

PHOTOASSIMILATE UTILIZATION DURING ANAEROBIC STRESS

An analysis of eight major crops produced in the USA indicated that there is a large unrealized genetic potential for plant yield, especially if these plant types were better adapted to the physiochemical environments in which they are grown (Boyer, 1982). Although genetic selection for adaptation to adverse environments has contributed significantly to increased agricultural production, the fundamental mechanisms of root tolerance to soil stresses have been rarely understood. It may be an understatement to suggest that plant roots which consume so much of the plant C have received too little attention by the scientific community.

Elucidation of the biological mechanisms responsible for plant adaptations to diverse environments is important but likely to be complex. As we develop generation after generation of predictive simulation models, our knowledge gained through these efforts should provide excellent opportunities for expanding agricultural production to more of the marginal soils. Despite this complexity, predicted to require approximately 2000-fold the current worldwide research efforts just to sequence the genome of complex forms of life (Blattner, 1983), the nominal cost of investigating plant root adaptation to adverse soil environments has the potential for greatly increasing crop productivity.

Historically, secondary assessments have been used to evaluate plant tolerance to adverse environments. Although, shoot to root ratios, yield, and other morphological parameters reflect the relative sensitivity of a plant to environmental stress, little new information can be gained by these studies unless there are concurrent assessments of the physiological or biochemical mechanisms which are controlled genetically. The mechanistic approach to our understanding of plant adaptation to stress environments would conserve our resources by enhancing nutrient and water acquisition, avoidance of toxic ions and disease, conservation of limited supplies of plant C, growth regulation, and many other attributes of plant tolerance.

Soil flooding or the large number of ubiquitous anaerobic microsites, estimated to approach 30% of the total pore space (Smith, 1976) of many well-drained agricultural soils, denotes an environmental limitation to the maximum genetic expression of most major crops. Crawford (1982) reported that despite extensive research on the flooding injury of agricultural crops, it is difficult to generalize on the initial causes of flooding damage to unadapted plants. Since there are many physiological responses to flooded conditions resulting in tissue collapse, those species which apparently have either the metabolic capacity to shunt the flow of electrons through the anaerobic tissue without accumulating large quantities of toxic anaerobic metabolites (e.g., acetaldehyde, ethanol, lactate, etc.) or the morphological adaptations which improve the oxygen supply to anaerobic tissue (e.g., disrupted tissue and the formation of aerenchyma pores), tolerate long- and short-term soil anaerobic stress. Since this chapter emphasizes the adverse effects of anaerobiosis on C use by crop plants, readers interested in the specifics of tolerance to anaerobiosis are referred to the excellent reviews by Jackson et al. (1982), Crawford (1982), and Hook and Crawford (1978).

Anaerobic metabolism of C compounds in the roots of stressed plants always decreases the efficient use of photoassimilates. A comparison of aerobic and anaerobic respiration indicates that oxidatively phosphorylated metabolic energy, which produces 300 Kcal for each glucose molecule, is reduced to 16 Kcal when oxygen is absent (Russell, 1977). This incomplete catabolism of substrates results in the production of many reduced organic compounds (Crawford, 1982). Some of these anaerobic metabolites are reported to be toxic. The accumulation of ethanol and perhaps other reduced compounds appears to be toxic to plant tissues as they modify cellular membranes at high concentrations (Chin and Goldstein, 1977; Rubin and Rottenberg, 1982). Accumulations of ethanol results in plasmolysis of root tip cells of dry edible beans subjected to anaerobiosis for 24 to 48 h (Smucker et al., 1978). Ethanol appears to relax plasma membranes by interfering with the sequestration (active transport) of Ca (Garrett and Ross, 1983) making them more permeable by altering their fluidity similar to that outlined in Fig. 2. Since most cell membranes become more rigid when subjected to an adverse environment, the metabolic administration of endogenous ethanol or perhaps other products of anaerobic metabo-

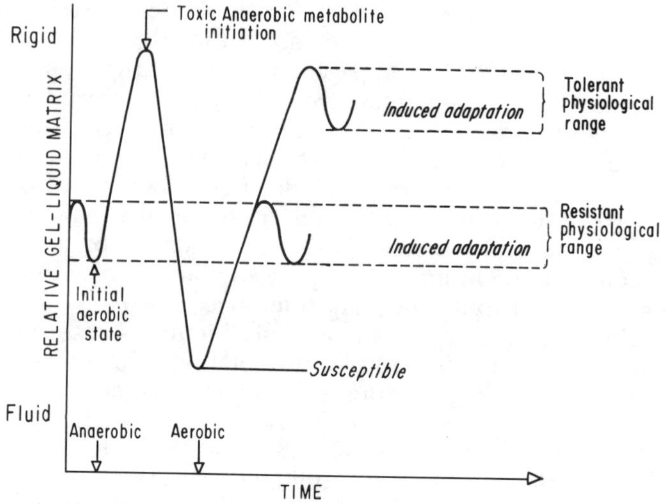

Fig. 2. Diagram of root plasma membrane responses to short-term anaerobic soil conditions. Suggested membrane fluidity responses to anaerobic metabolite accumulations are summarized before, during and following anaerobic soil conditions.

lism, appear to temporarily relax the membrane returning the gel-liquid matrix to a relatively more functional state. The net accumulation rate (i.e., production minus excretion) of ethanol appears to influence the relative tolerance of adapted or unadapted plants to soil flooding (Crawford, 1978). Once stressed membranes are aerated via soil drainage or the formation of gas pores in plant tissue (Jackson et al., 1982) the matrix of plant membranes may remain fluid (susceptible cultivars), return to the former state (resistant cultivars) or achieve a more rigid matrix (characteristic of conditioned and more tolerant cultivars) (Fig. 2).

It has also been reported that maintenance of the gel-liquid structure of human erythrocyte membranes depends upon glycolytic metabolism (Haest and Deuticke, 1975). Their report indicates that membrane asymmetry and fluidity were directly related to the rate of glycolysis. Applying this mechanism to root membranes suggests that most stress-induced metabolic changes alterning glycolytic metabolism could result in membrane dysfunction (e.g., leakiness, cytoplasmic pH changes, excessive respiratory consumption of soluble carbohydrates, etc.). Assuming that the hypothesis outlined in Fig. 2 has some validity, both the toxic anaerobic metabolite accumulation hypothesis of Crawford (1978) and the tissue modification hypothesis of Jackson et al. (1982) are two of the many mechanisms participating in the complex process of root adaptation to anaerobic soil conditions.

Regardless of the mechanisms involved in plant tolerance to anaerobic stress, it has been reported that roots subjected to these stresses lose more C (Rovira, 1969; Hale et al., 1971; Smucker and Erickson, 1976). Although ethanol and acetaldehyde production is usually associated with anaerobic stress, these products also accumulate in plant tissue subjected to other types of stress (Kimmerer and Kozlowski, 1982). Crawford (1982) reported that more than 17 hydrocarbons are accumulated and may be exuded by plants subjected to anaerobic conditions. The loss of these compounds represents a substantial C drain from already stressed plants. Two dry edible bean cultivars cultured in aseptic mist chambers (Smucker and Erickson, 1976) for 17 days and labeled with $^{14}CO_2$, lost from 19 to 34% of their labeled compounds through the roots (Table 4). These data indicate that stressed bean plants have very leaky and actively respiring root systems which lose up to one-third of the soluble photoassimilates. Although there appears to be some genetic variability in the retention capacities of nonstressed roots, adverse environments minimize the genetic influence on root leakage as anaerobic stress increased the C losses by the roots of these two genotypes to 33 and 34%.

Carbon losses by flooded root systems appear to be related to the internal and external oxygen content, duration, and frequency of flooding, and the genotype. Roots of two dry edible bean cultivars exposed to anaerobic (N_2) conditions exuded ethanol into the surrounding nutrient solution at similar rates during the first stress period (Shadan, 1980). However, roots of the more tolerant 'San Fernando' cv. lost less ethanol to the surrounding media during the aerated and second anaerobic periods (Fig. 3). These data suggest greater production and/or loss by the less tolerant variety. Since no aerenchyma were observed in the stems of these cultivars (unpublished data) and the average root porosity was 3.4% (Schumacher and Smucker, 1981) which is well below the maximum root porosity of maize (*Zea mays* L.) (10%) or rice (*Oryza sativa* L.) (33%) (Luxmoore et al., 1970), the greater quantity of ethanol exuded by stressed 'Seafarer' roots may have resulted from the induced activity of the alcohol dehydrogenase (ADH) enzymes. Activity of this induced enzyme complex has been reported to be 12.5 and 1.7 p*M* ADH (mg protein)$^{-1}$ min^{-1} for Seafarer and San Fernando, respectively (Noor, 1979).

Table 4. Distribution of ^{14}C-label in 17-day-old *Phaseolus vulgaris* cultivars cultured in aseptic mist chambers and subjected to two 72-h periods of anaerobic conditions. Tops were continuously labeled with $^{14}CO_2$ from Day 6 to 17 and roots were subjected to N_2 gas from days 6 to 9 and 13 to 15. Values are expressed as percent of total label in the plant and mist chamber system. Replicated experiments were repeated once (Shadan, 1980).

Treatment	Plant		Soluble root exudates	Respired root exudates
	DPM × 10^6† gdw^{-1}		%	
Aerobic control				
Seafarer	10.59	81.3	0.14	18.6
San Fernando	6.68	75.1	0.09	24.8
Anaerobic control				
Seafarer	8.21	66.7	1.5	31.8
San Fernando	6.92	65.7	1.3	33.0

† DPM represents the total ^{15}C activity expressed as disintegrations/minute, gdw = Gram dry weight.

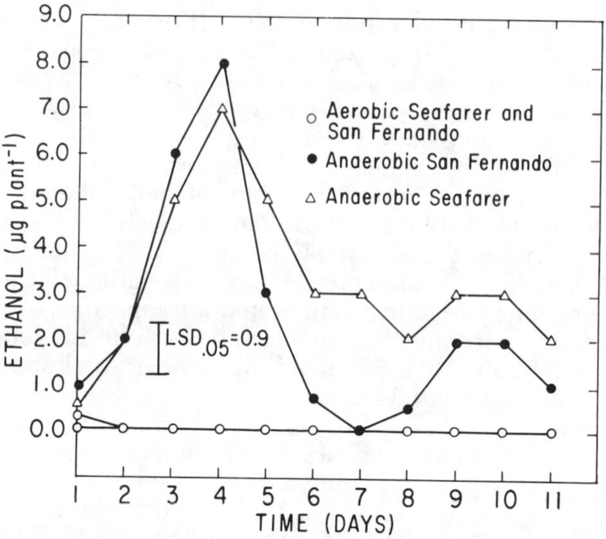

Fig. 3. Ethanol accumulations in the sterile mist chambers of 6-day-old dry edible bean root systems subjected to 72-h periods of anaerobic (N_2) conditions on Days 1 and 7.

Most of the ethanol and perhaps other toxic metabolites produced by roots grown under anaerobic conditions appear to be lost to the surrounding medium with only a small portion transported via the xylem to the shoot (Smucker and Adler, 1980). In greenhouse experiments, Seafarer plants were grown in aerated and nonaerated solutions containing various levels of ethanol. Ethanol concentrations of the nutrient solution and accumulations in the xylem were measured (Fig. 4). After 48 h, most of the ethanol added to the aerated controls was absorbed and metabolized with essentially no accumulation in the xylem exudates. However, four times more ethanol remained in the nutrient solution of the anaerobically treated roots when 125 mg ethanol L^{-1} was added, and approximately 30 mg of ethanol g^{-1} of root fresh weight accumulated in the xylem exudate. These data suggest that for dry edible beans, root anaerobiosis stimulates the production of ethanol, presumably by inducing the ADH enzyme complex (Crawford, 1967). Ethanol is disseminated from the root primarily by exudation or is metabolized and transported to the shoot.

The most damaging effects of ethanol and other C losses to stressed root systems may be the combined attractant and stimulative effects these C sources have on motile zoospores of pathogenic microorganisms (Young et al., 1977) and fungal growth (Smucker and Erickson, 1976; Cannell and Jackson, 1981). The cause and effect relationships between the organic compounds of root exudates and plant disease have been summarized by Schroth and Hildebrand (1964). It is generally accepted that plant roots exude sufficient quantities of organic compounds to

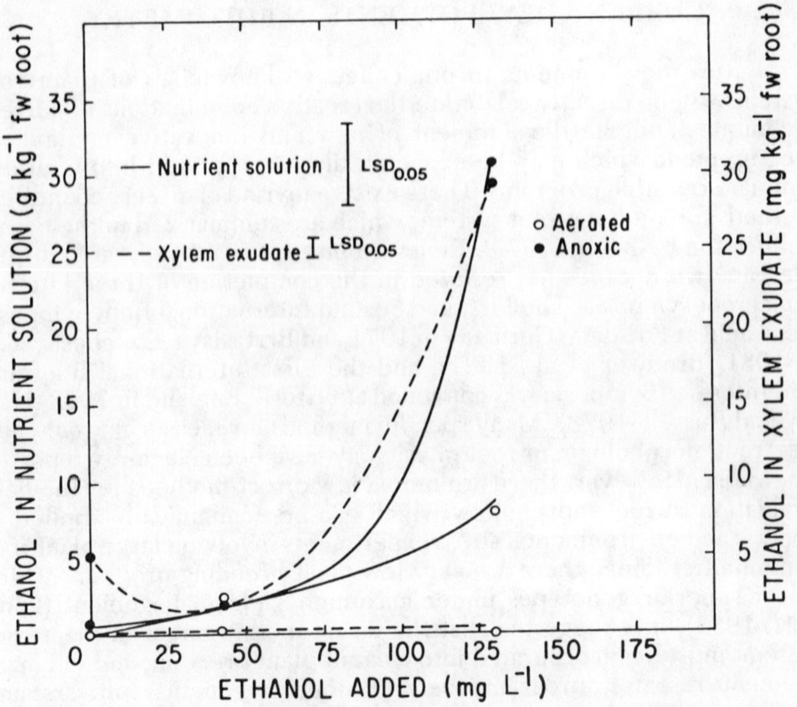

Fig. 4. Influence of ethanol concentrations added to the nutrient solution on the ethanol accumulation in xylem exudates of dry edible beans and the disappearance of ethanol from the solution after a treatment period of 48 h.

support large populations of microorganisms in the rhizosphere and that growth of certain microorganisms are influenced by the quality of these root exudates. Cochrane et al. (1963) reported that conidial germination of the root pathogen *Fusarium solani* F. sp. *phaseoli* required exogenous C and N as well as a growth factor from yeast which could be replaced by ethanol, acetaldehyde, or one of several amino acids. The synergistic effects of environmental stress and root disease hypothesized by Smucker and Erickson (1976) have been confirmed for the *Fusarium* root rot complex of dry edible beans grown on compacted soils (Burk et al., 1980). These reports suggest that those response mechanisms controlling the disease tolerance of host plants and the pathogenicity of causative organisms appear to be modified by the degree of environmental stress. These and other ecological relationships of the rhizosphere are essentially unknown. Additional interdisciplinary research among the abiotic and biotic stress disciplines of plant science and microbiol ecology is necessary before we will be able to elucidate with confidence the complex interactions which control the dynamic root and soil interface.

CURRENT DEVELOPMENTS IN RHIZOMETRY

Future developments in our collective knowledge of plant root systems will be directly related to the creative combinations of current techniques and the development of new and innovative methods of measurement which may be economically incorporated into multidisciplinary research programs. There exists a myriad of direct and indirect methods for studying root systems which are summarized in the timely publication by Bohm (1979). Renewed interest in root research during the past two decades has resulted in the completion of three international root symposia. The First and Second International Root Symposia were held at Potsdam, Germany in 1971 and Brataslava, Czechoslovakia in 1981 (Brouwer et al., 1981); and the First International Root and Soil Interface Symposia was convened at Oxford, England in 1978 (Harley and Russell, 1978). Many specific methods for measuring root ecology, root morphology, or root physiology have been elegantly reported elsewhere. However, there are only a few direct methods for the field evaluation of root morphology which can be economically applied to genetic and environmental stress experiments involving large plant root communities. Since there are very few plant breeding programs which select superior genotypes under maximum yield management (Lambert, 1983), it is suggested that the direct measurement of plant root systems must be incorporated into current plant breeding and soil management research programs before superior genetic cultivars and maximum-yield management systems will be developed.

Recent developments in the economical and quantitative evaluation of spacial and temporal root development have the potential for partially resolving the dilemma of expensive root research. Combinations of the field installation of many minirhizotron tubes with current state-of-the-art video microcameras developed by Sanders and Brown (1978) and improved by Upchurch and Ritchie (1983) have tremendous potential for evaluating in situ growth for many plant and soil communities. Although the initial investment for equipment ranges from \$15 000 to \$20 000, the manual and technical costs of operation are minimal (personal communication from Dr. Joe Ritchie). An additional problem with this approach is the unknown reason for the absence of roots in the surface few centimeters of soil above minirhizotrons, installed at angles which are less than perpendicular to the soil surface.

The direct method of extracting root and soil samples from field experiments throughout the growing season is laborous. Yet, if the roots are quantitatively separated from the soil, this approach provides excellent information. Destructive sampling procedures have been accepted in the past and, I believe, will continue to be the best approach for evaluating real interactions between plant roots and their soil environments. This rationale prompted the development of a root and soil profile sampler (Srivastava et al., 1982) and a rapid and quantitative method for separating roots and other organic debris from mineral soils (Smucker et al., 1982). The hydropneumatic elutriation manifold system provides a relatively inexpensive apparatus which improves the efficiency for recovering roots from mineral soils, reported to be 60 to 80% (Martin and Puckridge, 1982), to greater than 99%. This system also removes operator errors, is equally efficient for all types of mineral soils and preserves the original morphology of most root samples. Root and soil profiles of large field experiments can be sampled and the roots quantitatively separated from the soil in 2 to 4 days by the procedure outlined in Fig. 5.

Currently, with the application of these methods, the greatest impediment to our root research programs is the qualitative analyses of randomly branched root morphology. Although computer aided measurements of root length have been reported (Voorhees et al., 1980; Costigan et al., 1982), there is a real need to establish second generation image processor rhizometers which will discriminate between plant roots and accompanying organic debris as well as measure the viable and nonviable morphological components of roots which have been quantitatively extracted from field experiments. The combination of direct methods of extraction and quantitative recovery with innovative methods of analyses will certainly assist plant and soil scientists in delineating the combined and separate influences of genetics and the environment on the development and maintenance of plant root systems.

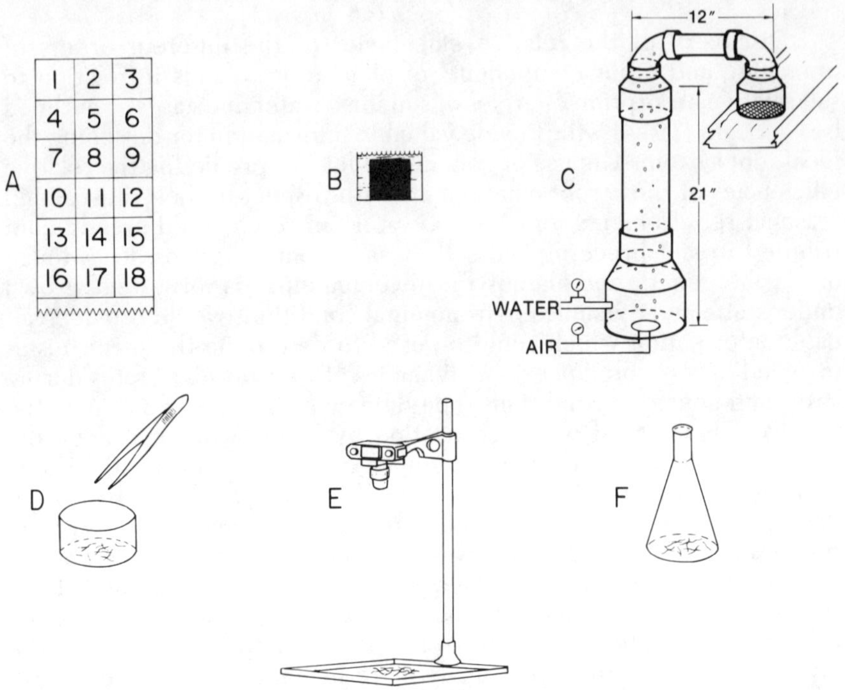

Fig. 5. Summary of the rapid and quantitative root and soil profile sampling procedure (A), sample preparation (B), root extraction by the hydropneumatic elutriation system (C), and sample analyses (D-F) of field-grown root communities.

RESEARCH IMPERATIVES

Establishment of a quantitatively based root analysis component to many of our plant and soil research programs has great economic and production potential. It is essential that the most efficient C balance between roots and shoots be determined for specific environmental conditions. The dynamics of C utilization in the rhizosphere, root respiration pathways, and the interactions of inorganic ions and hormones which influence root branching are essentially unknown. Carbon losses by plant roots result in complex associations of root and soil microorganisms which contribute both positively and negatively to the efficiency of the whole plant. It appears reasonable to assume that plant root systems could be developed to retain greater proportions of their C compounds for the benefit of the host.

Since root development can be altered by fertilizer placement and improved soil physical conditions, additional soil and plant management systems should be designed to investigate the C efficiency and plant yields. Especially plants having smaller root systems early in the growing season which develop deep and narrow root profiles as the season progresses. Genetic control of root branching, membrane elasticity of root cells, enzyme induction, hormone ratios, and the ideal C and N balance of whole plant systems are essentially unknown for most crop species.

Discovery of the relative efficiencies of the different orders of branching and other components of plant root systems in relation to the relative absorption energies of soil ions, water and gases as outlined by Clarkson (1974) will provide valuable information for designing the ideal root system. The use of systems models for predicting the relative efficiencies of these root components with respect to their diurnal and seasonal requirements for C, ions, water, air, etc. would greatly contribute to plant breeding and soil management programs. Knowledge of the ontogenetic and phenotypic developments of root systems grown under static and dynamic environmental conditions would provide valuable information which could be used to determine the mechanisms involved in root tolerance. The dynamics of root turnover rates during stress and nonstress conditions should be investigated to determine the relative efficiency of water absorption by viable and physiologically dead roots. Since a substantial number of reports indicate that dead roots absorb water, more research should be directed toward the relationship between plant or leaf water potential and the absorption surfaces of viable and dead roots. The specific photoassimilate requirements, cytokinin production, ion absorption, water absorption, C leakage, and respiration rates for each order of root branching during development, maintenance, death, and disappearance should also be investigated. Until these and many more biochemical, morphological, and physiological details of the plant root are understood, little progress will be made in genetically designing a C-efficient root system for specific soil types and specific levels of soil management.

SUMMARY AND CONCLUSIONS

Past morphological investigations of plant root systems have contributed to our knowledge of root responses to soil environments. Reports of C utilization by plant roots and the influence of stress environments on the allocation of C have been summarized to emphasize another component of the complex root-soil interface. Our knowledge of the biochemical and physiological responses of roots to specific environmental conditions has been limited by the paucity of standardized measurements. Many semiquantitative yet expedient techniques used to measure the morphological root responses of genotypes to specific soil environments are incomplete and in many cases highly inaccurate. Therefore, it is necessary to develop new and quantitative methods for measuring ontogenetic developments of root systems so that the mechanisms controlling root tolerance may be more completely defined. The development of superior cultivars and high yielding management systems will be greatly curtailed until these mechanisms are better understood.

REFERENCES

1. Arkin, G.F., and H.M. Taylor. 1981. Modifying the root environment to reduce crop stress. ASAE Monogr. no. 4. American Society of Agricultural Engineering. St. Joseph, Mich.
2. Barber, D.A., and K.B. Gunn. 1974. The effect of mechanical forces on the exudation of organic substances by the roots of cereal plants grown under sterile conditions. New Phytol. 73:39–45.
3. ----, and J.K. Martin. 1976. The release of organic substances by cereal roots into soil. New Phytol. 76:69–80.
4. Blattner, F.R. 1983. Biological frontiers. Science 222:719–720.
5. Bohm, W. 1979. Methods of studying root systems. Springer-Verlag, New York.
6. Boyer, J.S. 1982. Plant productivity and environment. Science 218:443–448.
7. ----, R.R. Johnson, and S.G. Saupe. 1980. Afternoon water deficits and grain yields in old and new soybean cultivars. Agron. J. 72:981–986.
8. Brouwer, R., O. Gasparikova, J. Kolek, and B.C. Loughman. 1981. Structure and function of plant roots. *In* Proc. of the 2nd Int. Symposium, Bratislava, Czech. Martinus Nijhoff, The Hague.
9. Burke, D.W., D.E. Miller, and A.W. Barker. 1980. Effects of soil temperature on growth of beans in relation to soil compaction and Fusarium root rot. Phytopathology 70:1047–1049.
10. Cannell, R.Q., and M.B. Jackson. 1981. Alleviating aeration stresses. p. 141–192. *In* H.M. Taylor and G.F. Arkin (ed.) Modifying the root environment to reduce crop stresses. American Society of Agricultural Engineering, St. Joseph, MI.
11. Carr, D.J., and J.S. Pate. 1967. Aspects of the biology of aging. Symp. Soc. Exp. Biol. 21:559–562.
12. Carson, E.W. 1974. The plant root and its environment. University Press of Virginia, Charlottesville.
13. Chatterton, N.J., and J.E. Silvius. 1979. Photosynthetic partitioning into starch in soybean leaves. Plant Physiol. 64:749–753.
14. Chin, J.H., and D.B. Goldstein. 1977. Effects of low concentrations of ethanol on the fluidity of spin-labeled erythrocyte and brain membranes. Mol. Pharmacol. 13:435–441.

15. Clarkson, D.T. 1974. Ion transport and cell structure in plants. John Wiley & Sons, New York.

16. Cochrane, J.C. et al. 1963. Spore germination and carbon metabolism in *Fusarium solani*. I. Requirements for spore germination. Phytopathology 53:1155–1160.

17. Coleman, D.C. 1976. A review of root production processes and their influence on soil biota in terrestrial ecosystems. p. 417–434. *In* J.M. Anderson and A. Macfodyen (ed.) The role of terrestrial and aquatic organisms in decomposition processes. Blackwell Scientific Publications, Oxford.

18. Costigan, P.A., J.A. Rose, and T. McBurney. 1982. A micro-computer based method for the rapid and detailed measurement of seedling root systems. Plant Soil 69:305–309.

19. Crawford, R.M.M. 1978. Metabolic adaptations to anoxia. p. 119–136. *In* D.D. Hook and R.M.M. Crawford (ed.) Plant life in anaerobic environments. Ann Arbor Science, Ann Arbor, MI.

20. ----. 1967. Alcohol dehydrogenase activity in relation to flooding tolerance in roots. J. Exp. Bot. 18:458–464.

21. ----. 1982. Physiological responses to flooding. p. 453–477. *In* Encyclopedia of plant physiology. New Series, Vol. 12B. Physiological plant ecology II.

22. de Wit, C.T. et al. 1978. Simulation of assimulation, respiration and transpiration of crops. John Wiley & Sons, New York.

23. Dixon, R.O.D., Y.M. Berlier, and P.A. Lespinat. 1981. Respiration and nitrogen fixation in nodulated roots of soybean and pea. Plant Soil 61:135–143.

24. Garrett, K.M., and D.H. Ross. 1983. Effects of in vivo ethanol administration on Ca^{2+}/Mg^{2+} ATPase and ATP-dependent Ca^{2+} uptake activity in synaptosomal membranes. Neurochem. Res. 8:1013–1028.

25. Geiger, D.R., and R.T. Giaquinta. 1982. Translocation of photosynthate. p. 345–386. *In* Govindjee (ed.) Photosynthesis. Vol. II. Development, carbon metabolism, and plant productivity. Academic Press, New York.

26. Haest, C.W.M., and B. Deuticke. 1975. Experimental alteration of phospholipid-protein interactions within the human erythrocyte membrane. Biochim. and Biophys. Acta 401:468–480.

27. Hale, M.G., C.L. Foy, and F.J. Shay. 1971. Factors affecting root exudation. Adv. Agron. 23:89–109.

28. Hansen, G.K., and C.R. Jensen. 1977. Growth and maintenance respiration in whole plants, tops and roots of *Lolium multiflorum*. Physiol. Plant 39:155–164.

29. Harley, J.L., and R.S. Russell. 1978. The root soil interface. *In* Proc. of the 1st Int. Root-Soil Interface Symp., Oxford. Academic Press, New York.

30. Hesketh, J.D., J.T. Woolley, and D.B. Peters. 1982. Predicting photosynthesis. p. 387–418. *In* Govindjee (ed.) Photosynthesis. Vol. II. Development, carbon metabolism, and plant productivity. Academic Press, New York.

31. Hook, D.D., and R.M.M. Crawford. 1978. Plant life in anaerobic environments. Ann Arbor Science, Ann Arbor, MI.

32. Jackson, M.B., B. Herman, and A. Goodenough. 1982. An examination of the importance of ethanol in causing injury to flooded plants. Plant Cell Environ. 5:163–172.

33. Johnen, B.G., and D.R. Sauerbeck. 1977. A tracer technique for measuring growth, mass and microbial breakdown of plant roots during vegetation. Soil organisms as components of ecosystems. Ecol. Bull. Stockholm 25:366–373.

34. Kimmerer, T.W., and T.T. Kozlowski. 1982. Ethylene, ethane, acetaldehyde, and ethanol production by plants under stress. Plant Physiol. 69:840–847.

35. Kuiper, Daan. 1983. Genetic differation in *Plantago major*: Growth and root respiration and their role in phenotypic adaptation. Physiol. Plant 57:222–230.

36. Lambers, H., D.A. Day, and J. Azcon-Bieto. 1983. Cyanide-resistant respiration in roots and leaves. Measurements with intact tissues and isolated mitochondria. Physiol. Plant 58:148–154.

37. Lambert, R.J. 1983. Hybrid selection under maximum yield management. p. 20. *In* Better Crops With Plant Food. Fall.

38. Lang, A., and M.R. Thorpe. 1984. Analyzing partitioning in plants. Plant Cell Environ. (In press).

39. Laties, G.G. 1982. The cyanide-resistant alternative path in higher plant respiration. Ann. Rev. Plant Physiol. 33:519–555.

40. Luxmoore, R.J., L.H. Stolzy, and J. Letey. 1970. Oxygen diffusion in the soil plant system. II. Respiration rate, permeability and porosity of consecutive excised segments of maize and rice roots. Agron. J. 62:322–324.

41. Martin, J.K. 1977. Factors influencing the loss of carbon from wheat roots. Soil Biol. Biochem. 9:1–7.

42. ----, and D.W. Puckridge. 1982. Carbon flow through the rhizosphere of wheat crops in south Australia. p. 77–82. *In* J.R. Freney and I.E. Galbally (ed.) Cycling of carbon, nitrogen, sulfur and phosphorus in terrestrial and aquatic ecosystems. Springer-Verlag, New York.

43. McCree, K.J. 1970. An equation for the rate of respiration of white clover plants grown under controlled conditions. p. 221–229. *In* I. Setlik (ed.) Prediction and measurement of photosynthetic activity. PUDOC Wageningen.

44. ----. 1982. The role of respiration in crop production. Iowa State J. Res. 56(3):291–306.

45. Minchin, F.R., and J.S. Pate. 1973. The carbon balance of a legume and the functional economy of its root nodules. J. Exp. Bot. 24:259–271.

46. Moldau, H., and A. Karolin. 1977. Effect of reserve pool on the relationship between respiration and photosynthesis. Photosynthetica 11:38–47.

47. Noor, N.M. 1979. Enzyme activity and ethanol accumulation as related to zinc uptake in four varieties of dry beans (*Phaseolus vulgaris* L.) under aeration stress. M.S. thesis, Michigan State Univ., East Lansing.

48. Pate, J.S., D.B. Layzell, and D.L. McNiel. 1979. Modeling the transport and utilization of carbon and nitrogen in a nodulated legume. Plant Physiol. 63:730–737.

49. Pavlychenko, T.K. 1937. Quantitative study of the entire root system of weed and crop plants under field conditions. Ecology 18:62–79.

50. Prikryl, Z., and V. Vancura. 1980. Root exudates of plant: VI. Wheat root exudation as dependent on growth, concentration gradient of exudates and the presence of bacteria. Plant Soil 57:69–83.

51. Rovira, A.D. 1969. Diffusion of carbon compounds away from wheat roots. Aust. J. Biol. Sci. 22:1287–1290.

52. Ruben, E., and H. Rottenberg. 1982. Ethanol-induced injury and adaptation in biological membranes. Fed. Proc. 41:2465–2471.

53. Russell, R.S. 1977. Plant root systems, their function and interaction with the soil. McGraw-Hill, New York.

54. Sanders, J.L., and D.A. Brown. 1978. A new fiber optic technique for measuring root growth of soybeans under field conditions. Agron. J. 70:1073–1076.

55. Sauerbeck, D.R., and B.G. Johnen. 1977. Root formation and decomposition during plant growth. p. 141–148. *In* Soil organic matter studies. 1976. Vol. 1. IAEA, Vienna.

56. Schroth, M.N., and D.C. Hildebrand. 1964. Influence of root exudates on root infecting fungi. Ann. Rev. Phytopathol. 2:101–132.

57. Schumacher, T.E., and A.J.M. Smucker. 1981. Mechanical impedance effects on oxygen uptake and porosity of dry bean roots. Agron. J. 73:51–55.

58. ----, and ----. 1984a. Carbon transport and root respiration responses to *Phaseolus vulgaris* genotypes to localized anoxia. (Submitted for publication).

59. ----, and ----. 1984b. Ion uptake and respiration of dry bean roots subjected to localized anoxia. (Submitted for publication).

60. Shadan, M.M. 1980. Fixation, translocation and root exudation of ^{14}C-labeled assimilates by two genotypes of *Phaseolus vulgaris* L. subjected to root anaerobiosis. M.S. thesis. Michigan State Univ., East Lansing.

61. Smith, A.M. 1976. Ethylene production by bacteria in reduced microsites and some implications to agriculture. Soil Biol. Biochem. 8:293–298.

62. Smucker, A.J.M., and F. Adler. 1980. Accumulation and loss of toxic assimilates by plant roots. Agron. Abstr. 72:93.

63. ----, and A.E. Erickson. 1976. An aseptic mist chamber system: A method for measuring root processes of peas. Agron. J. 68:59–62.

64. ----, B.D. Knezek, and G.R. Hooper. 1979. Influence of short-term oxygen stress upon the translocation of zinc in navy beans (*Phaseolus vulgaris* L.) p. 434. *In* J.L. Harley and R.S. Russell (ed.) The soil-root interface. Academic Press, New York.

65. ----, S.L. McBurney, and A.K. Srivastava. 1982. Quantitative separation of roots from compacted soil profiles by the hydropneumatic elutriation system. Agron. J. 74:500–503.

66. Srivastava, A.K., A.J.M. Smucker, and S.L. McBurney. 1982. A mechanical sampling method of multiple soil-plant root studies. Trans. ASAE 25:868–871.

67. St. Jon, T.V., and D.C. Coleman. 1983. The role of mycorrhizae in plant ecology. Can. J. Bot. 61:1005–1014.

68. Taylor, H.M., and B. Klepper. 1978. The role of rooting characteristics in the supply of water to plants. Adv. Agron. 30:99–128.

69. Thorpe, M.R., A. Lang, and P.E.H. Minchin. 1983. Short term interactions between flows of photosynthate. J. Exp. Bot. 34:10–19.

70. Torrey, J.G., and D.T. Clarkson. 1975. The development and function of roots. Academic Press, New York.

71. Upchurch, D.R., and J.T. Ritchie. 1983. Root observations using a video recording system in minirhizotrons. Agron. J. 75:1009–1015.

72. Voorhees, W.B., V.A. Carlson, and E.A. Hallauer. 1980. Root length measurement with a computer-controlled digital scanning microdensitometer. Agron. J. 72:847–851.

73. Warembourg, F.R., and G. Billes. 1979. Estimating carbon transfers in the plant rhizosphere. p. 183. *In* J.L. Harley and R.S. Russell. (ed.) The soil-root interface. Academic Press, New York.

74. Young, B.R., F.J. Newhook, and R.N. Allen. 1977. Ethanol in the rhizosphere of seedlings of *Lupinus angustifolius* L. N. Z. J. Bot. 15:189–191.

Chapter 3

Water Flux in the Soil-Root Continuum[1]

MORRIS G. HUCK[2]

The living tissues of a growing crop may contain more than 90% water, yet under normal conditions this water can be lost to the atmosphere in only a few minutes of transpiration. The requirement of growing plants for continuous access to water to replace that lost by transpiration, can easily be demonstrated by cutting a leaf in bright sunlight. Wilting will be observed within a very few minutes.

Transpiring leaves obtain water from their associated root systems which, in turn, extract water from reserves held in the surrounding soil. The concept of a Soil-Plant-Air Continuum (SPAC) (Philip, 1966) was also described by Huber (1924), Burch et al. (1978), and others. It provides a means for describing water flow by equations of the same general form as those used in electrical engineering (Fig. 1). Although not exactly analogous (Smith, 1983), comparisons between soil water flow through the SPAC and the flow of electricity through a resistance network have enhanced our understanding of the system in recent years (Molz, 1976; Hillel, 1977, 1980).

To assure its xylem transport system access to an adequate water supply, a plant must maintain a vigorous root system in contact with soil moisture reserves throughout its functional life. The root system,

---

[1] Contribution from ARS, USDA, in cooperation with Dep. of Agronomy and Soils, Alabama Agric. Exp. Stn., Auburn University, AL 36849.

[2] Soil scientist, ARS, USDA, Auburn, AL 36849.

Copyright © 1984 Soil Science Society of America, Crop Science Society of America, and American Society of Agronomy, 677 South Segoe Road, Madison, WI 53711. *Roots, Nutrient and Water Influx, and Plant Growth.*

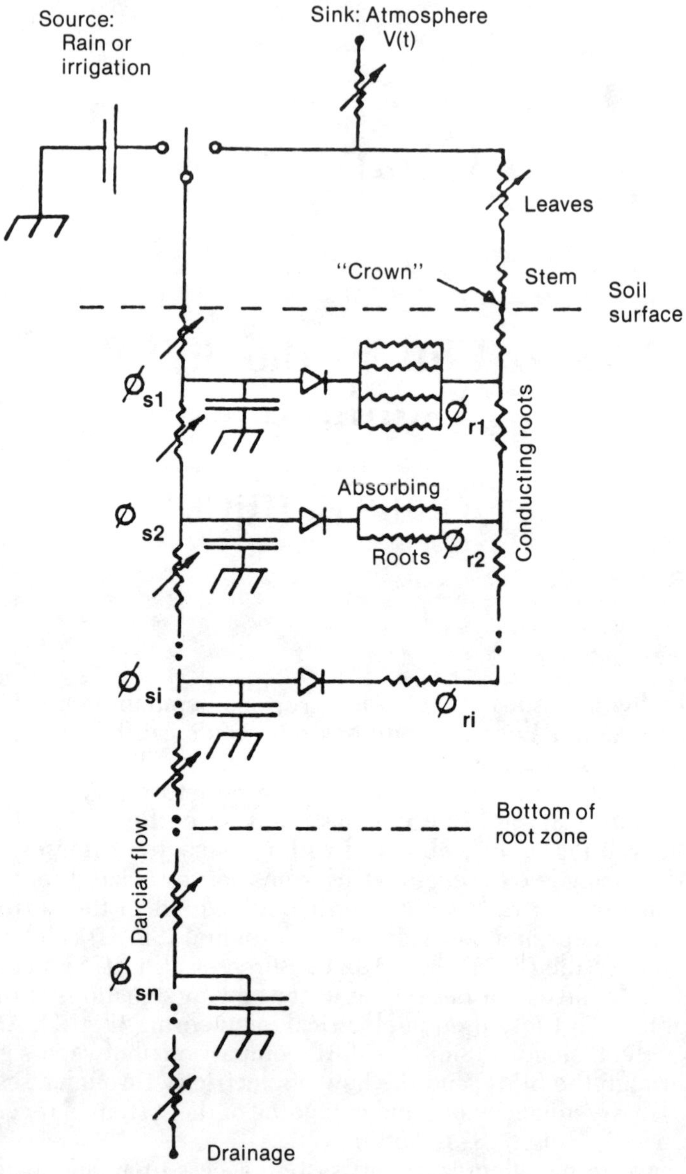

Fig. 1. (after Hillel, 1977): Schematic representation of a root system in layered soil viewed as a resistance network. Water held in capacitors moves through variable resistances. Diodes permit flow into the roots but not from roots into the soil.

in turn, obtains sugars and other organic substances required for its growth and metabolism from leaves, as described by Huck and Hillel (1983). Water reaching the leaves also contains inorganic salts, selected from those dissolved in the soil solution, by the root system. Computation of the time integrals of relevant biochemical processes and thus, growth can be related to specific aspects of the water balance (Boyer, 1982).

The total water requirement for a crop growing over a complete season is generally greater than the amount of soil water that can be held in storage. These reserves must be periodically replenished by rainfall or irrigation. Figure 2 illustrates the relevant macroscopic flow processes. Water enters the system by rainfall or irrigation, and is held in storage until removal, either by drainage to groundwater, direct evaporation from the soil surface, or by root uptake to meet the transpiration requirements of the growing plants. Losses to the atmosphere by transpiration and direct evaporation are mutually exclusive, with the proportional contribution of each being influenced by canopy shading of the soil surface (a function of leaf area index). The relationship between water storage capacity of most agricultural soils and the rate of water usage by growing crops is somewhat analogous to the relationship between water stored in plant tissue and the rate of water uptake by roots.

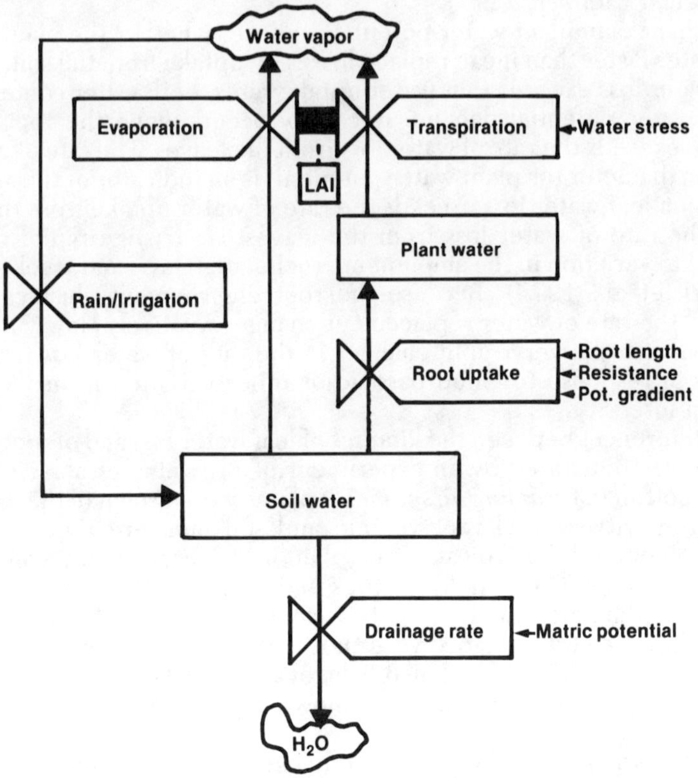

Fig. 2. (after Huck and Hillel, 1983) Schematic diagram illustrating water flow in the soil-root continuum. The "valve" symbols represent water flow rates between rectangular storage compartments or to external "sinks" in the atmosphere or groundwater.

Since stomatal closure impedes both CO_2 uptake and water loss from the leaves, the amount of photosynthetic C assimilated by plants whose stomata close in midmorning to conserve water will be significantly less than that of control plants having root systems with access to an adequate water supply (Ishihara, 1979). Whenever stomata open to admit CO_2, water vapor escapes to the atmosphere. Even partial dehydration of the leaves induces stomatal closure, thus limiting photosynthesis when light and temperature conditions would otherwise permit maximum photosynthetic rates. Measurements of diurnal net photosynthetic assimilation rates with root systems under differing moisture regimes (Turner et al., 1978; Huck et al., 1983) support this concept.

Because of this requirement for continuous water availability, water flux through the soil-root continuum (from soil, through roots, and into the leaves) is a very practical criterion for evaluating plant water status. The instantaneous value of plant water potential (a function of tissue water content) can be computed by integrating the difference between the rate of water loss (associated with photosynthesis and transpiration) and the water uptake rate (associated with functions of the root system) (Huck and Klepper, 1977).

Changes in plant water potential indicate whether the plant is losing water faster than it can replace losses by uptake from the soil. When leaf water loss exceeds root water uptake rates, both water content and plant water potential decline. Recovery occurs when the root water uptake exceeds the rate of water loss from the leaves. The rate of change in stem diameter (or plant water potential) is an indicator of the amount by which leaf water loss exceeds the rate of water uptake from the soil.

The rate of water loss from the leaves can change rapidly in response to variation in the ambient microclimate (Davis and Huck, 1978; Stansell et al., 1974). Because the root environment changes more slowly, the rate of water replacement changes relatively slowly in comparison with the very rapid changes in the rate of water loss from the leaves in response to cloud passage or other variation in incident radiation intensity.

Differences between the kinetics of leaf water loss and of root water uptake are illustrated by an experiment of Browning et al. (1975), in which cotton (*Gossypium hirsutum* L.) plants were grown in the Auburn rhizotron with several root volume and soil moisture regimes. Leaf water potential for a typical cotton plant in this experiment, computed by the dynamic flux method of Huck and Klepper (1977) from continuous stem diameter measurements, is shown in Fig. 3. The soil profiles beneath these cotton plants were thoroughly wet by heavy rainfall on 30 and 31 July 1972. Less than 0.5 cm of additional rain fell throughout the month of August, so the data shown represent a single extended drying cycle.

Details of measured stem diameter changes and corresponding water potential values as well as the pressure chamber water potential measurements used to calibrate the continuous water potential estimates, are shown for two selected 3-day intervals in Fig. 4. From these figures,

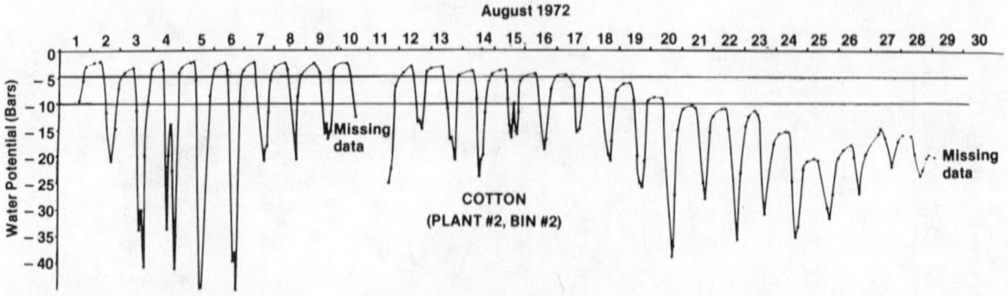

Fig. 3. Plant water potential as a continuous function of time computed for a typical plant from the experiments of Browning et al. (1975) by the dynamic flux method of Huck and Klepper (1977).

it can be seen that the cotton plant grew rapidly during early stages of the drying cycle (6 to 8 August, Fig. 4a). Leaf water potential returned to values above −5 bars soon after sunset. Later in the drying cycle, as soil water reserves were depleted (23 to 25 August, shown in Fig. 4b), the stem diameter (water potential) measurements indicated a much slower rate of recovery after sunset. The summary plot (Fig. 3) indicates that after 18 August nightly recovery from the leaf water stress of the preceding day was so slow that equilibrium water potential (top line, −5 bars) was not attained by sunrise.

The importance of water flux through the soil-root continuum, as demonstrated by these experiments, lies chiefly in its effect upon plant water potential, which in turn influences growth and overall plant productivity. A postulated mechanism illustrating how the kinetics of water flow can influence plant growth is shown in Fig. 5. When stomata open and transpiration begins at sunrise, the plant gradually comes under increasing water stress, represented by a reduction in plant water potential. Soil water potential changes relatively slowly; reduced plant water potentials therefore increase the gradient between leaves and the various components along the flow pathway back through the root system to the water held in soil storage (Burch et al., 1978; Hillel, 1980). Although intermediate values along the flow pathways are not well established (Fiscus, 1972), the general shape of the potential gradient shown in Fig. 6 is probably realistic. As long as flow resistance remains nearly constant, the rate of water flow through the root-soil continuum will increase as leaf water potential drops.

If a plant has many roots actively absorbing water, it can transpire rapidly for extended periods of time and the stomata will not be required to close for water conservation. With increasing plant water requirement, however, the rate of water flux across the soil-root continuum will eventually exceed the maximum allowable value (Lang and Gardner, 1970) for more and more individual rootlets. Thus, an ever increasing demand will be placed upon the remaining roots, until finally the aggregate root system will no longer be able to meet the transpiration requirement of fully opened stomata; closure is then necessary to avoid desiccation and tissue injury.

Faiz and Weatherly (1982) proposed that root shrinkage reported by Huck et al. (1970), occurs as the water flux into an individual root

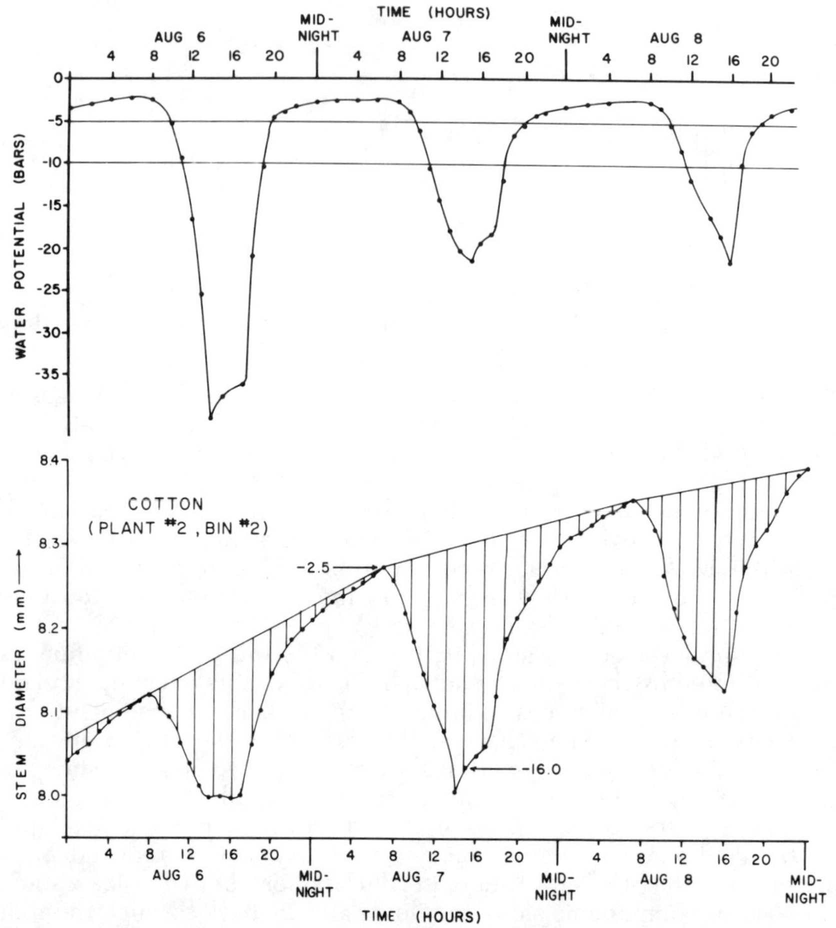

Fig. 4a. Detail of cotton water potential from Fig. 3 for 6 to 8 August when soil water reserves were adequate to meet evaporative demand. Numbers (with arrows pointing to line) are pressure-chamber water potential values measured at the times shown. Lower curve illustrates method of computation from stem diameter measurement (Huck and Klepper, 1977).

exceeds the critical velocity for soil in that root's immediate vicinity. As the critical flux rate (potential gradient) is exceeded for some roots, the demand for water from each remaining root increases further. Finally, when there are no longer enough roots in contact with soil water reserves to meet the transpiration requirement, plant water potential approaches its wilting point equilibrium (lower boundary line in Fig. 5). Stomata closure then begins as a safety measure, to retard further water loss and prevent death or injury to the delicate living tissues of the leaves.

The approach to wilting point equilibrium is not a function of water loss rate alone, however. It is also a function of root water uptake rate or, more properly, the difference between both rates, summed over

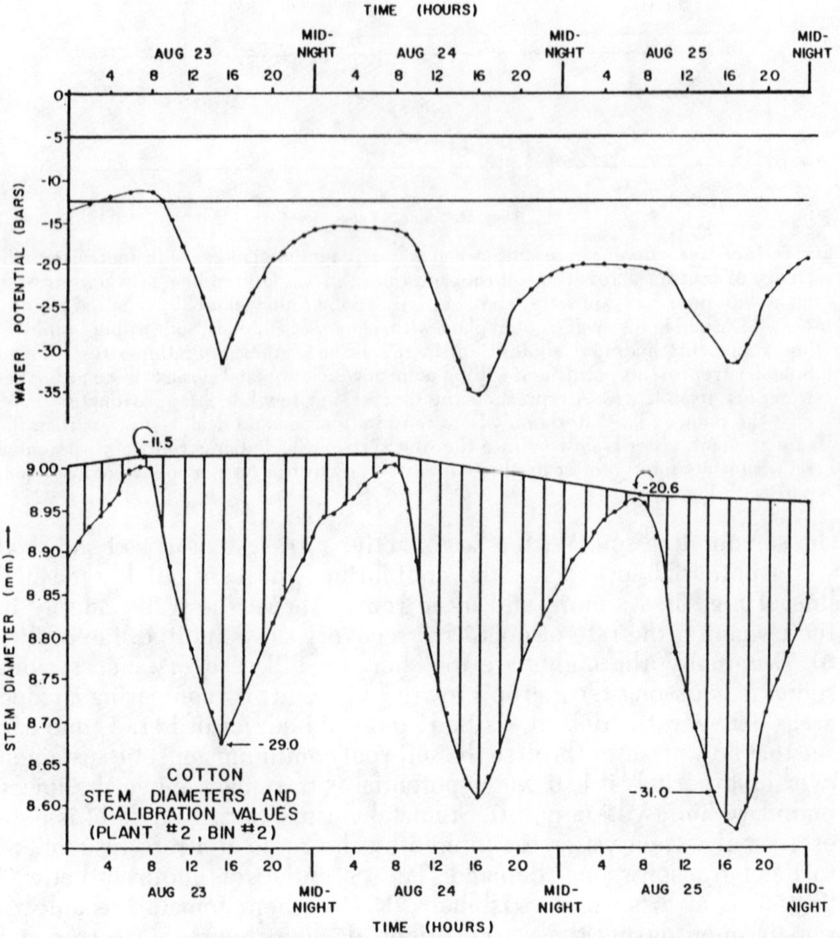

Fig. 4b. Detail of cotton leaf water potential from Fig. 3 for 23 to 25 August when root contact with soil water reserves was inadequate to meet evaporative demand of the shoot (Data from Browning et al., 1975).

time. For sustained photosynthetic CO_2 fixation, the plant must transpire rapidly for extended periods of time without needing to restrict stomatal aperture for water conservation.

The recovery phase, when plant water potential asymptotically approaches its hydrated equilibrium value, is shown in Fig. 5 as the curve from point B to A′. With the onset of darkness, the leaves begin to rehydrate. This reverse side of the SPAC water balance equilibrium is more closely related to the ability of root systems to transport water into the leaves, while the declining phase (discussed earlier) is more responsive to variation in the rate of water loss from the leaves (Davis and Huck, 1978).

The time required for rehydration is largely independent of climate, but it varies with the ability of root systems to obtain water from

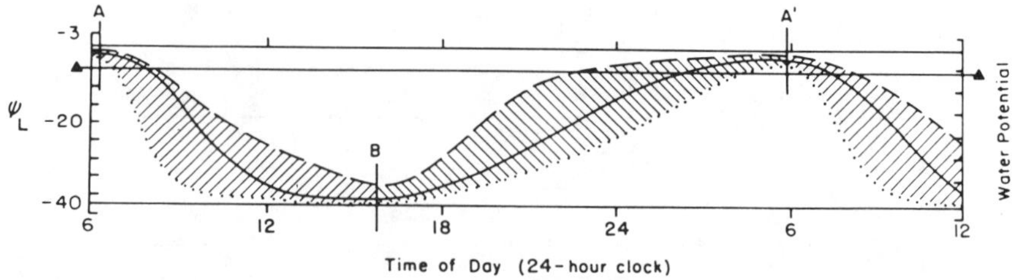

Fig. 5. Idealized curves representing leaf water potential changes with increasing efficiency of transfer across the soil-root interface. .... Dotted line = water stressed plant with poor root-soil water contact; ______ Solid line = normally-adapted plant; ------ Dashed line = well-rooted plant with many roots in moist soil. Upper boundary line represents hydrated equilibrium in the absence of transpiration stress; Lower boundary represents permanent wilting point at which stomatal closure presents further water loss from leaves. A represents the time of sunrise while B (approximately 1700 h) is the point in late afternoon when transpiration demand declines to less than the rate at which water is flowing into the root system. Shaded area represents potential reduction in photosynthetic productivity caused by restricted root-soil water flux (compare with Fig. 4).

the surrounding soil. With a large, active root system in wet soil, leaf water potential approaches the equilibrium line very quickly (dashed line of Fig. 5). As more and more roots exhaust the water supply in their vicinity, the rate of nocturnal recovery slows (dotted line of Fig. 5). Eventually, the nights are too short for full recovery, and serious reductions in long-term crop growth can result. By comparing shaded areas between the dotted, dashed, and solid curves of Fig. 5, one can see that flow of water through the soil-root continuum must be sustained over long periods if leaf water potential is to remain above the lower boundary line (wilting point). Stomatal closure during daylight is necessary only when water is supplied by the roots at a rate insufficient to meet transpirational demand. The potential for photosynthetic C fixation is at a maximum (Ishihara, 1979) when stomata are able to remain open during the warm, bright, daylight hours. Since stomatal closure impedes CO_2 uptake as well as water loss from the leaves (deWit and Goudriaan, 1974), it is obvious that when stomata must close in midmorning to conserve water, the rate of photosynthetic C assimilation will be reduced in comparison to that measured in control plants having root systems with access to an adequate water supply.

Long-term trends showing the effect of restricted rooting volume on plant growth are demonstrated by the experiments of Klepper et al. (1973), in which rooting volume and irrigation practices were varied, providing a direct comparison of root system effectiveness in differing soil environments. Additional data are given for each treatment by Browning et al. (1975). As each soil profile dried over a period of several days, more and more roots experienced difficulty reaching adequate water soil water reserves and the rate of nocturnal recovery slowed. Huck et al. (1983) showed a similar effect when the growth habits of soybean (*Glycine max* L.) root systems were altered by imposition of water stress over periods of several days.

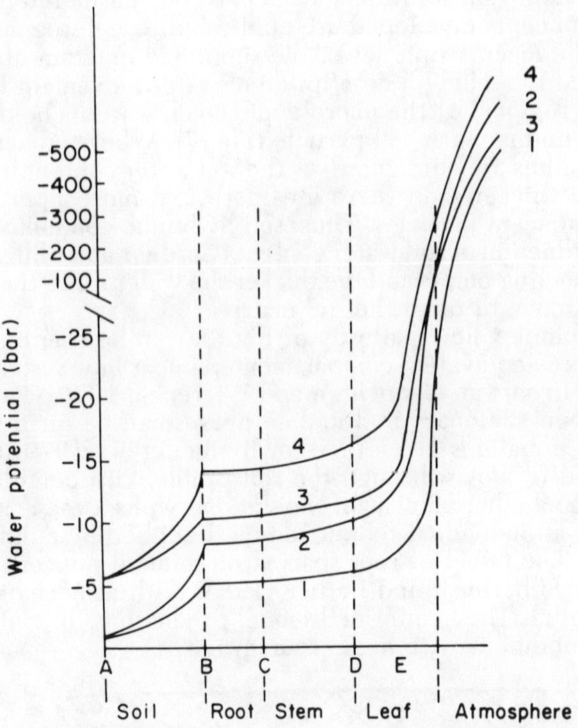

Fig. 6. (after Hillel, 1980). Idealized water potential gradients through the SPAC (figure not drawn to scale, and shown to illustrate general principles only). Curves 1 and 2 represent plant water removal from relatively wet soil at low and high transpiration rates, respectively; curves 3 and 4 illustrate relative shapes of water potential gradients through the Soil-Plant-Air Continuum (SPAC) at low and high transpiration rates, respectively, after soil water potential has been reduced to −6 bars. (Note reversal of positive direction on y-axis between Fig. 3–5 vs. Fig. 6, due to differing conventions in original publications.)

Short-term reductions in shoot growth and C-fixation rates have been observed long before the soil has dried to the extent discussed above. Boyer (1970) suggested that water potentials below −3 to −5 bars may have inhibited protein synthesis in maize (*Zea mays* L.) plants, while soybean photosynthesis continued until water potential dropped below −11 bars. In addition to stomatal closure, other metabolic processes (McPherson and Boyer, 1977; Thomas and Andre, 1982) can also be disrupted at low leaf water potentials. It is tempting to speculate that protein synthesis (and growth of new plant tissue) occurred in the cotton plants of Fig. 4 on the nights of 6 to 8 August, but by 23 to 25 August the water potential was too low for protein synthesis at any time during the night. Thus, the plant would be unable to regenerate the enzymatic components needed for sustained growth and development.

Water flow through the soil-root continuum can be studied at two levels: a macroscopic and a microscopic scale. For a practical understanding of soil-root water flux, each must be considered in such a way that the concepts developed are applicable to heterogenous, layered soils. At the macroscopic level, the simplified diagram of Fig. 2 must be modified to include a description of water movement between different soil regions. At the microscopic level, water is held as capillary films surrounding each soil particle (Fig. 7). Where adjacent particles touch, the films are contiguous; as the soil water content increases, the films grow thicker, and there are more and more points of contact between adjacent particles. Thus, soil hydraulic conductivity is an exponential function of soil water content (Gardner and Ehlig, 1962). The more connecting points and the thicker the water films, the more easily water can move through the soil matrix.

The complex flow pathway of Fig. 7a can be simplified and represented as an equivalent compartmented linear flow system illustrated in Fig. 7B (from van Genuchten and Wierenga, 1976). Partitioning of ions between stationary and mobile phases can be quantitatively described by equations given there or by Rao et al. (1979). Soluble nutrients tend to move through the soil profile with percolating water; ions having a higher affinity for the stationary phase move more slowly. Activity of an individual rootlet can be described by a different set of equations. The effect of root spacing on mineral nutrient uptake was studied by Silberbush and Barber (1983) with a mechanistic mathematical model. They confirmed model predictions in controlled environment conditions with high accuracy.

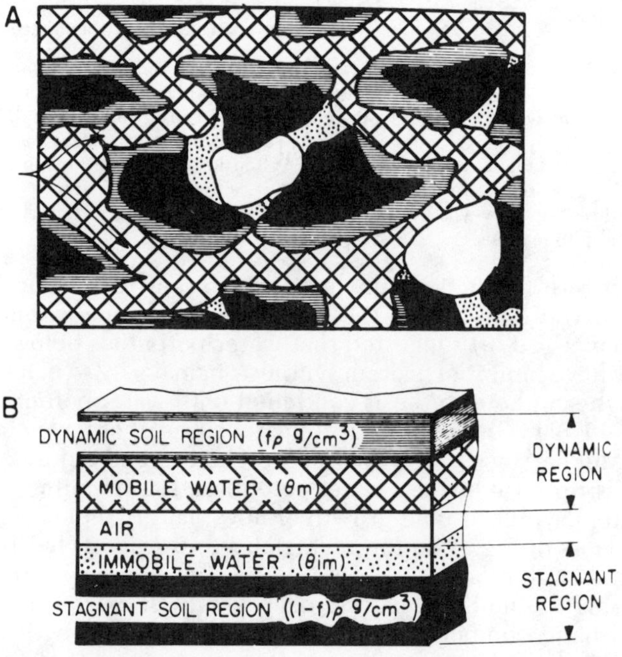

Fig. 7. (after van Genuchten and Wierenga, 1976). Soil particles are shown surrounded by films of capillary water and connecting irregular pore spaces. (A) actual geometry; (B) simplified model. The shading patterns in (A) and (B) represent the same regions.

An early single-root water uptake model named TROIKA was described by Lambert and Penning de Vries (1973) for considering the kinetics of water flow from soil through the root and into the atmosphere. The model predicted results similar to those of Molz (1975) who evaluated water flow toward roots in a cylindrical coordinate system with a set of equations that generated a family of parametric curves. Gardner (1960) made a number of simplifications and reached similar conclusions more than 15 years earlier. When the TROIKA model was extended to include varying lengths of root in a heterogeneous soil matrix (Huck and Graves, 1976), the need for additional assumptions in a more complex macroscopic system became apparent.

More complex computer simulation models such as that described by Huck and Hillel (1983) permit analysis of a one-dimensional system describing vertical water flow between adjacent horizontal layers and the interactions with growing plants as illustrated in Fig. 8. In this system, water is added to the soil surface by rainfall or irrigation and moves down through successive "layers" of soil under the influence of gravity. Because of the exponential relationship between water content and hydraulic conductivity, water flows rapidly through the wet soil near the surface until it encounters a boundary, or "wetting front" separating wet from dry soil (Green and Ampt, 1911), or a compacted layer, or a change in texture. As more and more water enters the soil, the wetting front progresses downward until it either establishes contact with another layer of soil wet from a previous rain or continues drainage into the groundwater. More recently, Philip (1966) and Swartzendruber and Hillel (1973) have comprehensively reviewed the subject of infiltration, while Whisler et al. (1982) described a two-dimensional simulation model that includes the effects of other microenvironmental variables such as soil aeration or mechanical impedance to root growth.

The total pore space between soil particles is generally constant (except after tillage or compaction operations, which alter the physical structure of the soil), so water entering the soil profile must displace the air that had filled the pore spaces of dry soil. At extremely high application rates, incoming water may trap air in the deeper soil pores, which will temporarily retard penetration of the wetting front. When rain or irrigation is finished, water will continue draining internally until the hydraulic potential is equal at all points (Hillel, 1980).

In the absence of crop cover, water will evaporate directly from the wet soil surface (Whisler et al., 1968). As the top layer of soil dries, capillary movement of water between deeper soil layers and the drier layers at the surface layers will be restricted by the exponential conductivity/water content relationship until at some point, capillary films are broken and further water movement is only by vapor transfer (Jackson et al., 1974; Fiscus and Huck, 1972). Thus, dry soil at the surface forms a mulch, restricting surface evaporation of water held in deeper soil layers until removal either by drainage to groundwater, by direct evaporation from the soil surface, or by root uptake to meet the transpiration requirements of growing plants. Losses to the atmosphere by

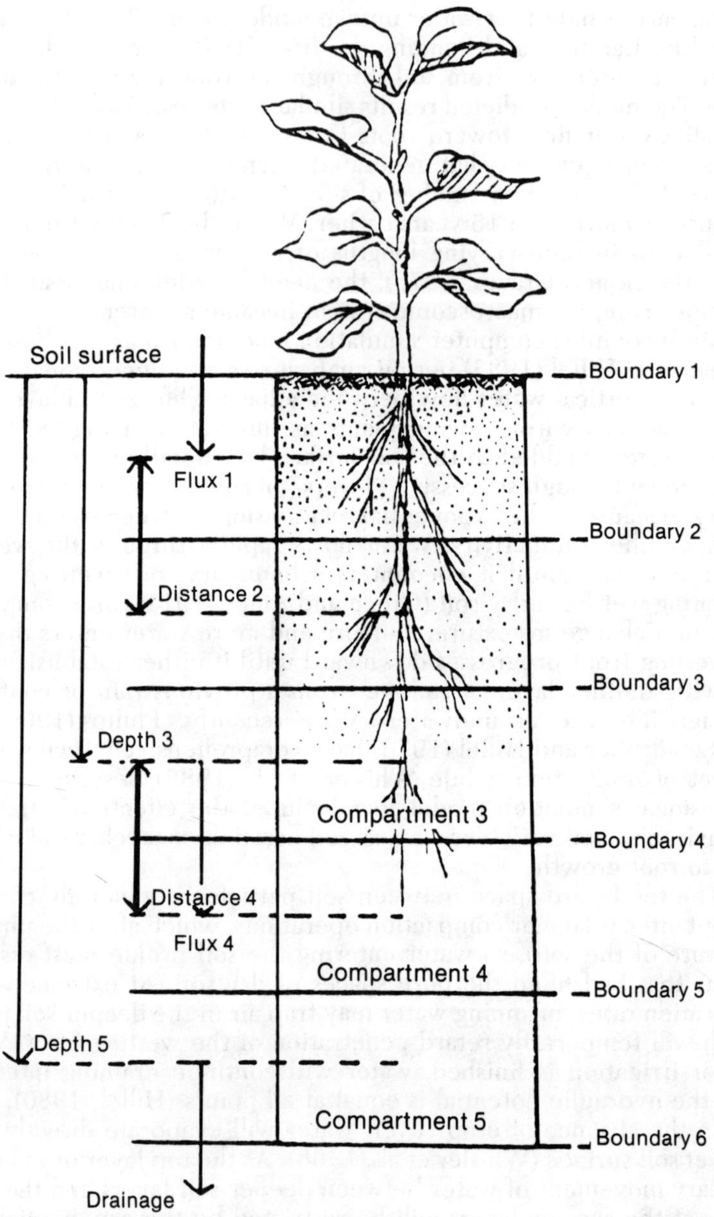

Fig. 8. Hypothetical plant growing in a one-dimensional soil consisting of uniform layers in the horizontal direction. Water flows within the soil body as described by Huck and Hillel (1983), or is removed by the roots penetrating each layer to replace water lost from the leaves by transpiration.

transpiration and direct evaporation are mutually exclusive, with the proportion controlled by canopy shading of the soil surface (a function of Leaf Area Index, LAI).

The interrelated patterns of soil water distribution and root distribution within the soil profile are continually changing (Huck, 1977). As plants grow, their roots penetrate into successively deeper regions of the soil (Fig. 8), removing water held there to satisfy the transpirational requirement of their leaves. Transpiration accounts for an increasing fraction of the total water lost from the soil profile as LAI increases (Huck and Hillel, 1983). When the canopy cover is incomplete, direct evaporation of water from the bare (unshaded) soil surface must also be considered. Soil moisture evaporation and plant transpiration (directly linked to root uptake rate) are complementary processes. Each is limited by the availability of water and solar energy, but direct evaporation is limited both by the degree of surface exposure and the ability of surface soil layers to conduct water held in deeper layers.

The spatial pattern of water uptake from various regions of the soil depends upon root distribution as well as the distribution of soil moisture within the soil profile (Taylor and Klepper, 1978). Water is removed from deeper layers as root penetration increases and water remaining near the surface is held at more negative matric potential. If the soil is uniformly wet, the root water uptake pattern is influenced mainly by root distribution and root resistance to longitudinal flow (Hillel, 1977). When moisture has been removed from soil regions initially having the greatest concentration of roots, the "center of gravity" of water uptake must shift to wetter soil layers. Because of their immature vascular systems, growing root tips must expand by taking up moisture from the immediately surrounding soil layers. Root proliferation is, therefore, much more rapid in soil with an adequate water supply than in drier regions of the soil (Burch et al., 1978). Sharp and Davies (1979) provided experimental evidence for an increase in root growth relative to shoot growth under water-stressed conditions. Since the life expectancy of small feeder roots is only a few days, root growth will appear to "track" regions of moist soil (Huck, 1977), with new roots forming in wetter regions adjacent to older roots while rootlets in dry soil die, and are not regenerated until the next rainfall or irrigation cycle.

When an adequate quantitative description of processes in the root-soil system is available, the equations can be combined in computer simulation models for optimization studies. Models such as those described by Baker et al. (1983), Jones et al. (1983), Huck and Hillel (1983), or Acock et al. (1984), for example, can assist in evaluating effects of alternative growth strategies. Ehlers et al. (1983) demonstrated that soils are far from homogeneous, and that our assumptions of uniformity are therefore questionable. A certain amount of extrapolation will always be required to bring error estimates within allowable limits.

After establishing confidence in a computer simulation model of root growth and function, one can ask specific questions with the model, obtaining simulated results, which can then be compared with experimental observation. Using the network model of Huck and Graves

(1976), for example, one might ask whether a plant with a longer, finer root system occupying the same soil volume (i.e., smaller root-root distances) would have less plant water stress than another with larger roots at a greater distance apart in the soil. Experiments by Fiscus and Markhart (1977) and Passioura (1972) found this to be true, but soil water reserves were depleted more rapidly in the process. The reason for this was anticipated by Gardner (1960, 1964) and further documented by Molz (1975, 1976). Not surprisingly, the simulation (which uses equations proposed by Gardner) also reaches the same conclusion.

A related question, somewhat similar to the first, is whether plants with 50% more root length on the same size shoot but distributed through a larger soil volume, would reduce plant water stress. Since it involves root growth, this question must be addressed with a more complex model, such as that of Huck and Hillel (1983) and the experimental data of Browning et al. (1975). The answer is again a conditional yes, depending on the water distribution. If roots can already reach the available water, there is little advantage to additional roots in the same soil region. When the existing root system is sparse and the new roots can explore soil not accessible to the preexisting roots, however, the additional roots would help (Hillel and Talpaz, 1976). Additional experimental data bearing upon this problem was reported by Huck et al. (1983).

The inverse question of whether plants can grow satisfactorily with only half their roots in moist soil and the other half in soil below the wilting point, is equally instructive. Unless rain or irrigation is imminent, roots in the dry soil contribute little. The data of Klepper et al. (1973), suggest that such plants may also grow satisfactorily, provided there are enough roots in the moist soil, and that the definition of "satisfactory" can be distinguished from that of "optimum." Total daily photosynthesis can be increased only when new roots grow into moist soils (roots in dry soil contribute little toward resupplying water lost from the leaves). Thus, with reference to Fig. 5, stomatal closure will occur earlier each day, and the recovery each night will be slower under these conditions.

A reduction in photosynthetic C fixation may have little effect upon economic return if crop growth is sink-limited (Korte et al., 1983; McMichael and Hesketh, 1982). As long as the stomata admit sufficient CO_2 to form the sugars needed to support growth of the desired crop components (e.g., seeds, or other storage organs), there is no economic advantage to fixing additional C. Growth of additional leaf and stem tissue may only compound harvesting problems, while contributing nothing to seed yield, for example. Thus, maximizing soil-water flux is not the same as maximizing economic return to the farmer (Boyer, 1982). The more general optimization question remains: the plant would be ahead to make an additional carbohydrate investment. This question requires exploration with a more complex model such as that of Huck and Hillel (1983) or Baker et al. (1983). The answer is largely dependent upon the initial and boundary conditions, which must be clearly specified.

Since root growth rates often exceed the rate of water movement through unsaturated soil, it may be advantageous for a plant to regenerate feeder roots in wet soil after each rainfall or irrigation event rather than to maintain a stationary root system throughout the rooting volume. The plant might be better off to discard older roots that have removed all the available water in a region rather than to maintain them until the next rain, which may not arrive until several more days have passed. Less metabolic C and energy may be required to regenerate another root later, when that volume of soil is rewet than would be expended on maintenance respiration (Penning de Vries, 1975) during the interval when the root apparently serves no useful function. Grasses generally adopt the "stake out" strategy, with a long-lived root system, whereas dicotyledonous crops (e.g., cotton and soybeans) tend toward the "commando" strategy of occupying a root volume a few days while the water lasts, and then moving into another region with large numbers of short-lived feeder roots.

Recent experimental reports, such as those of Reicosky et al. (1982) or Zur et al. (1982) suggest that hydraulic conductivity of soil in the immediate vicinity of the root may differ markedly from that of bulk soil. As more experimental data become available, alternative forms of the water extraction equations such as those proposed by Raats (1974) or Fiscus et al. (1983) can be substituted into the various models for comparison of prediction with measurement data.

REFERENCES

1. Acock, B., V.R. Reddy, F.D. Whisler, D.N. Baker, J.M. McKinion, H.F. Hodges, and K.J. Boote. 1984. The soybean simulator GLYCIM: Model documentation. 1982. Mississippi Agric. Exp. Stn. Bull. no. 0002 (In press). Starkville, MS.
2. Baker, D.N., J.R. Lambert, and J.M. McKinion. 1983. GOSSYM: A simulator of cotton crop growth and yield. South Carolina Exp. Stn. Bull. no. 1089, Clemson, SC.
3. Boyer, J.S. 1970. Differing sensitivity of photosynthesis to low leaf water potentials in corn and soybeans. Plant Physiol. 46:236–239.
4. ----. 1982. Plant productivity and environment. Science 218:443–448.
5. Browning, V.D., H.M. Taylor, M.G. Huck, and B. Klepper. 1975. Water relations of cotton: A rhizotron study. Auburn Univ. Exp. Stn. Bull. no. 467, p. 1–70.
6. Burch, G.J., R.E.G. Smith, and W.K. Mason. 1978. Agronomic and physiological responses of soybean and sorghum crops to water deficits. II. Crop evaporation, soil water depletion and root distribution. Aust. J. Plant Physiol. 5:169–177.
7. Davis, J.M., and M.G. Huck. 1978. Identifying turgor responses of water-stressed cotton to rapid changes in net radiation using spectral analysis. Crop Sci. 18:605–612.
8. de Wit, C.T., and J. Goudriaan. 1974. Simulation of ecological processes. PUDOC, Wageningen.
9. Ehlers, W., U. Köpke, F. Hesse, and W. Böhm. 1983. Penetration resistance and root growth of oats in tilled and untilled loess soil. Soil Tillage Res. 3:261–275.
10. Faiz, S.M.A., and P.E. Weatherly. 1982. Root contraction in transpiring plants. New Phytol. 92:333–343.
11. Fiscus, E.L. 1972. *In-situ* measurement of root-water potential. Plant Physiol. 50:191–193.

12. ----, and M.G. Huck. 1972. Diurnal fluctuations in soil water water potential. Plant Soil 37:197–202.
13. ----, A. Klute, and M.R. Kaufmann. 1983. An interpretation of some whole plant water transport phenomena. Plant Physiol. 71:810–817.
14. ----, and A.H. Markhart III. 1977. Relationships between root system water transport properties and plant size in *Phaseolus*. Plant Physiol. 64:770–773.
15. Gardner, W.R. 1960. Dynamic aspects of water availability to plants. Soil Sci. 89:63–73.
16. ----. 1964. Relation of root distribution to water uptake and availability. Agron. J. 56:41–45.
17. ----, and C.F. Ehlig. 1962. Some observations on the movement of water to plant roots. Agron. J. 54:453–456.
18. Green, W.H., and G.A. Ampt. 1911. Studies on soil physics I: Flow of air and water through soil. J. Agric. Sci. 4:1–24.
19. Hillel, D. 1977. Computer simulation of soil water dynamics. Int. Dev. Res. Centre, Ottawa, Canada.
20. ----. 1980. Applications of soil physics. Academic Press, New York.
21. ----, and H. Talpaz. 1976. Simulation of root growth and its effect on the pattern of soil water uptake by a non-uniform root system. Soil Sci. 121:307–312.
22. Huber, B. 1924. Die Beurteilung des Wasserhaushaltes der Pflanze. Ein Beitrag zur vergleichenden Physiologie. Jahrb. Wiss. Bot. 24:1–120.
23. Huck, M.G. 1977. Root distribution and water uptake patterns. p. 215–225. *In* J.K. Marshall (ed.) The belowground ecosystem: A synthesis of plant-associated processes. Colorado State Univ. Dep. Range Sci. Series no. 26, Ft. Collins.
24. ----, and J.E. Graves. 1976. Network circuit analysis of water flow in a heterogeneous time-varying root soil system. 1976 Summer Computer Simulation Conference, Washington, DC.
25. ----, and D. Hillel. 1983. A model of root growth and water uptake accounting for photosynthesis, respiration, transpiration, and soil hydraulics. p. 273–303. *In* D. Hillel (ed.) Advances in irrigation research. Vol. II. Academic Press, New York.
26. ----, K. Ishihara, C.M. Peterson, and T. Ushijima. 1983. Soybean adaptation to water stress at selected stages of growth. Plant Physiol. 73:422–427.
27. ----, and B. Klepper. 1977. Water relations of cotton. II. Continuous estimates of plant water potential from stem diameter measurements. Agron. J. 69:593–597.
28. ----, ----, and H.M. Taylor. 1970. Diurnal variation in root diameter. Plant Physiol. 45:529–530.
29. Ishihara, K. 1979. Diurnal course of stomatal aperture of leaf blades in rice plants. Jpn. Agric. Res. Q. 13:85–89.
30. Jackson, R.D., R.J. Reginato, B.A. Kimball, and F.S. Nakayama. 1974. Diurnal soil-water evaporation: Comparison of measured and calculated soil-water fluxes. Soil Sci. Soc. Am. Proc. 38:861–866.
31. Jones, J.W., B. Zur, and K.J. Boote. 1983. Field evaluation of a water relations model for soybean II: Diurnal fluctuations. Agron. J. 75:281–286.
32. Klepper, B., H.M. Taylor, M.G. Huck, and E.L. Fiscus. 1973. Water relations and growth of cotton in drying soil. Agron. J. 65:307–310.
33. Korte, L.L., J.H. Williams, J.E. Specht, and R.E. Sorensen. 1983. Irrigation of soybean genotypes during reproductive ontogeny I: Agronomic responses. Crop Sci. 23:521–527.
34. Lambert, J.R., and F.W.T. Penning de Vries. 1973. Dynamics of water in the soil-plant-atmosphere system: A model named TROIKA. p. 257–273. *In* B. Yaron et al. (ed.) Arid-zone irrigation. Springer-Verlag, New York.
35. Lang, A.R.G., and W.R. Gardner. 1970. Limitation to water flux from soils to plants. Agron. J. 62:693–695.
36. McMichael, B.L., and J.D. Hesketh. 1982. Field investigations of the response of cotton to water deficits. Field Crops Res. 5:319–333.

37. McPherson, H.G., and J.S. Boyer. 1977. Regulation of grain yield by photosynthesis in maize subjected to a water deficiency. Agron. J. 69:714–718.

38. Molz, F.J. 1975. Potential distributions in the soil-root system. Agron. J. 67:726–729.

39. ----. 1976. Water transport in the soil-root system: Transient analysis. Water Resour. Res. 12:805–807.

40. Passioura, J.B. 1972. The effect of root geometry on the yield of wheat growing on stored water. Aust. J. Agric. Res. 23:745–752.

41. Penning de Vries, F.W.T. 1975. The cost of maintenance processes in plant cells. Ann. Bot. 39:77–92.

42. Philip, J.R. 1966. Plant water relations: Some physical aspects. Ann. Rev. Plant Physiol. 17:245–268.

43. Raats, P.A.C. 1974. Steady flows of water and salt in uniform soil profiles with plant roots. Soil Sci. Soc. of Am. Proc. 38:717–722.

44. Rao, P.S.C., J.M. Davidson, R.E. Jessup, and H.M. Selim. 1979. Evaluation of conceptual models for describing nonequilibrium adsorption-desorption of pesticides during steady-flow in soils. Soil Sci. Soc. Am. J. 43:22–28.

45. Reicosky, D.C., T.C. Kaspar, and H.M. Taylor. 1982. Diurnal relationship between evapotranspiration and leaf water potential of field-grown soybeans. Agron. J. 74:667–673.

46. Sharp, R.E., and W.J. Davies. 1979. Solute regulation and growth by roots and shoots of water-stressed maize plants. Planta 147:43–49.

47. Silberbush, M., and S.A. Barber. 1983. Prediction of phosphorus and potassium uptake by soybeans with a mechanistic mathematical model. Soil Sci. Soc. Am. J. 47:262–265.

48. Smith, R.E. 1983. Approximate soil water movement by kinematic characteristics. Soil Sci. Soc. Am. J. 47:3–8.

49. Stansell, J.R., B. Klepper, V.D. Browning, and H.M. Taylor. 1974. Effect of root pruning on water relations and growth of cotton. Agron. J. 66:591–592.

50. Swartzendruber, D., and D. Hillel. 1973. The physics of infiltration. p. 3–15. *In* A. Hadas et al. (ed.) Physical aspects of soil water and salts in ecosystems. Springer-Verlag, New York.

51. Taylor, H.M., and B. Klepper. 1978. The role of rooting characteristics in the supply of water to plants. Adv. Agron. 30:99–128.

52. Thomas, D.A., and M. Andre. 1982. The response of oxygen and carbon dioxide exchanges and root activity to short-term water stress in soybean. J. Exp. Bot. 33:393–405.

53. Turner, N.C., J.E. Begg, H.M. Rawson, S.D. English, and A.B. Hearn. 1978. Agronomic and physiological responses of soybean and sorghum crops to water deficits III: Components of leaf water potential, leaf conductance, $^{14}CO_2$ photosynthesis, and adaptation to water deficits. Aust. J. Plant Physiol. 5:179–194.

54. van Genuchten, M. Th., and P.J. Wierenga. 1976. Mass transfer studies in sorbing porous media I. Analytical solutions. Soil Sci. Soc. Am. J. 40:472–477.

55. Whisler, F.D., A. Klute, and R.J. Millington. 1968. Analysis of steady-state evaporation from a soil column. Soil Sci. Soc. Am. Proc. 32:167–174.

56. ----, J.R. Lambert, and J.A. Landivar. 1982. Predicting tillage effects on cotton growth and yield. p. 179–198. *In* P.W. Unger et al. (ed.) Predicting tillage effects on soil physical properties and processes. Spec. Pub. no. 44, American Society of Agronomy and Soil Science Society of America, Madison, WI.

57. Zur, B., J.W. Jones, K.J. Boote, and L.C. Hammond. 1982. Total resistance to water flow in field soybeans: II. Limiting soil moisture. Agron. J. 74:99–105.

Chapter 4

Plant Root Morphology and Nutrient Uptake[1]

S. A. BARBER AND M. SILBERBUSH[2]

Root length and radius can influence nutrient uptake by plant roots growing in soil. Total nutrient uptake depends on root surface area, and the rate of increase and uptake per unit of root surface. Mean rate of nutrient uptake per unit of root surface depends both on the uptake kinetics of the root and the nutrient supply characteristics of the soil. The extent and nature of the root system can vary from the fine fibrous root system common to perennial grasses to the coarser roots found in some herbacious species. The significance of amount of root surface also depends on the nutrient required per unit of root.

Recent research on nutrient flux into plant roots growing in soil has greatly increased our understanding of the processes involved. It has also enabled us to evaluate the significance of each parameter in the system. In this chapter we use root surface area and root radius data to evaluate the significance of root morphology in determining nutrient uptake under a variety of nutrient-crop-soil situations. Our knowledge of nutrient uptake by plant roots has led to its mathematical description with mechanistic mathematical models (Nye et al., 1975; Claassen and Barber, 1976; Cushman, 1979). These models aid in un-

[1] Contribution from Dep. of Agronomy, Purdue Univ., West Lafayette, IN 47907.

[2] Professor of agronomy, Dep. of Agronomy, Purdue Univ., West Lafayette, IN 47907 and professor, Inst. for Desert Res., Univ. of the Negev, Beer-Sheva 84120, Israel.

Copyright © 1984 Soil Science Society of America, Crop Science Society of America, and American Society of Agronomy, 677 South Segoe Road, Madison, WI 53711. *Roots, Nutrient and Water Influx, and Plant Growth.*

derstanding the system as well as determining the significance of the parameters involved.

PREDICTING UPTAKE WITH MATHEMATICAL MODELS

Claassen and Barber (1976) developed a model that described nutrient uptake by roots growing in soil. The model described the extent of the root system and its increase with time by using values for initial root length, rate of increase in root length, and mean radius. Roots were assumed to be smooth cylinders, without interroot competition. The model gave a value for root surface area and its radial geometry as it changed with time. Nutrient uptake by the root was assumed to be a function of concentration in soil solution at the root surface as described by the Michaelis-Menten equation. Supply of nutrients to the root through the soil was assumed to be by mass-flow and diffusion. Cushman (1979) developed an improved model which also accounted for competition between adjacent roots for nutrients.

The Claassen-Barber model has 10 parameters, while the Cushman model has one additional parameter to account for root competition. The parameters are:

Root morphology.

1. L_o, initial root length, cm.
2. k, rate of root growth. This coefficient may be for a linear, exponential or logistic relation, depending on which describes root growth for the species and conditions used.
3. r_o, average root radius, cm.

Root uptake kinetics.

4. Imax, maximal influx at large values of C_l, nmol $m^{-2}s^{-1}$.
5. Km, value of C_l-Cmin where In, net influx, is ½Imax, mmol L^{-1}
6. Cmin, C_l, where In = 0, mmol L^{-1}.

Soil nutrient supply.

7. C_{li}, initial nutrient concentration in the soil solution, mmol L^{-1}.
8. b, buffer power, a mean value describing ΔC_s, changes in concentration of labile nutrient on the solid phase in mmol dm^{-3} with ΔC_l, changes in nutrient concentration in soil solution, mmol L^{-1} of soil solution, for the concentration range used.
9. D_e, effective diffusion coefficient, which describes the effective rate of diffusion of a nutrient per unit volume of bulk soil, cm^2s^{-1}.
10. v_o, the rate of water influx, cm s^{-1}.
11. r_1, mean half distance between root axes, cm. A root axis is a line through the center of the root cylinder.

Methods for measuring the parameters used in the model are described in Silberbush and Barber (1983a).

The basic transport equation used in the models to describe radial flux to the root by mass-flow and diffusion as affected by uptake by the root is:

$$\frac{\delta C_l}{\delta t} = \frac{1}{r}\frac{\delta}{\delta r}\left(r\, D_e \frac{\delta C_l}{\delta r} + \frac{v_o r_o C_l}{b}\right), \qquad [1]$$

where r is radial distance from the root axis, C_l is concentration of the nutrient in soil solution, and t is time of uptake.

An initial and either one or two boundary conditions are required for solution. The initial boundary condition of the equation is:

$$t = 0,\ r > r_o,\ C_l = C_{li}, \qquad [2]$$

where C_{li} is the initial nutrient concentration in the soil solution.

The first boundary condition is:

$$D_e b \frac{\delta C_l}{\delta r} + v_o C_l = \frac{Imax(C_l - Cmin)}{Km + C_l - Cmin} \text{ at } r = r_o,\ t > 0, \qquad [3]$$

where Cmin is C_l when In = 0 (In is net influx; nmol $m^{-2}s^{-1}$).

The second boundary condition, an addition used by Cushman to account for competition between adjacent roots, is:

$$D_e b \frac{\delta C_l}{\delta r} + v_o C_l = 0,\ r = r_1,\ t > 0, \qquad [4]$$

where r_1 is the mean half distance between root axes. If there is no competition $C_l = C_{li}$ at $r = r_1$ at $t > 0$.

As the root grows, uptake is calculated for each additional unit of root growth and total uptake is obtained by summarizing uptake for all root units. Hence, plant uptake is dependent both on the uptake of each root unit and the number of root units in the system.

The models, except where otherwise modified, assume ions move to the root by diffusion and mass-flow and that roots are smooth cylinders absorbing nutrients from the soil solution according to Michaelis-Menten kinetics. No effect of root exudates, microorganisms, root hairs, or mycorrhizae are assumed unless the model is modified to account for their effects. Detailed assumptions of the model are given by Barber and Cushman (1981).

MODEL VERIFICATION

The Claassen-Barber model was verified by Claassen and Barber (1976) for K uptake by corn (*Zea mays* L.) growing in four soils, one with five levels of K.

Potassium uptake predicted by the model is compared with observed K uptake in Fig. 1. An r^2 of 0.87 indicates a good correlation between the observed and predicted uptake, however, a slope of 1.566 indicates overprediction of uptake probably because the Claassen-Barber did not allow for interroot competition.

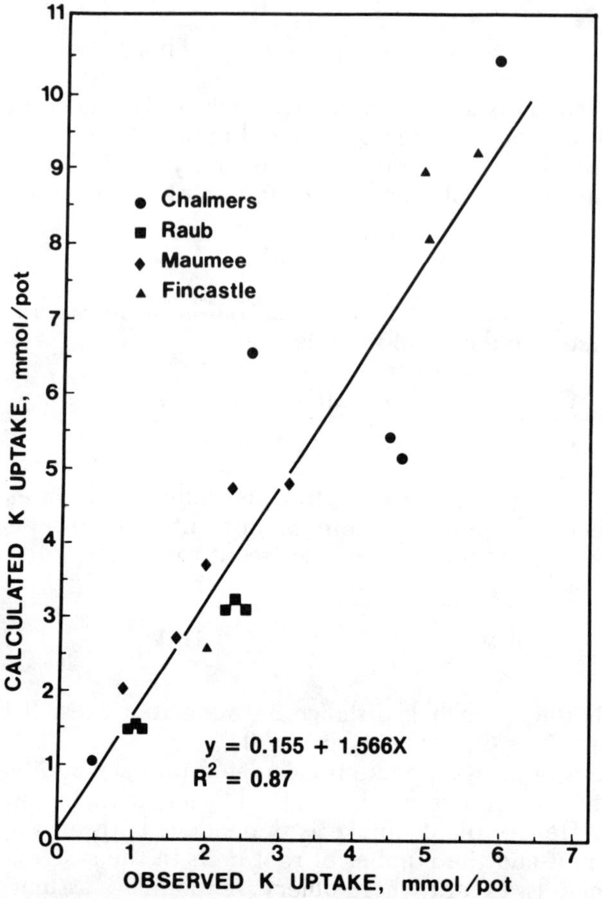

Fig. 1. Relation between predicted K uptake and observed K uptake by corn from four soils in 3-L pots in a growth chamber (Claassen and Barber, 1976).

Schenk and Barber (1979) verified the Claassen-Barber model for P uptake. They used six Indiana soils varying widely in organic matter and clay content. The correlation between observed and predicted P uptake of r = 0.93 (Fig. 2) indicated a close correlation and a slope of 1.09 indicated the observed values were close to the predicted values. Phosphorus diffuses much shorter distances than K so that competition between roots for P in this experiment was not likely to be a factor in determining uptake.

Schenk and Barber (1980) also verified the model for predicting P uptake by three corn single-cross hybrids growing on one soil in the field. The results shown in Fig. 3 give a reasonably close correlation (r = 0.91) between observed and predicted uptake. The slope of 0.67 indicated underprediction. Underprediction may have been due to uptake by root hairs.

Cushman (1979) developed a mechanistic model which also accounted for reduction in ion flux to the root due to competition between

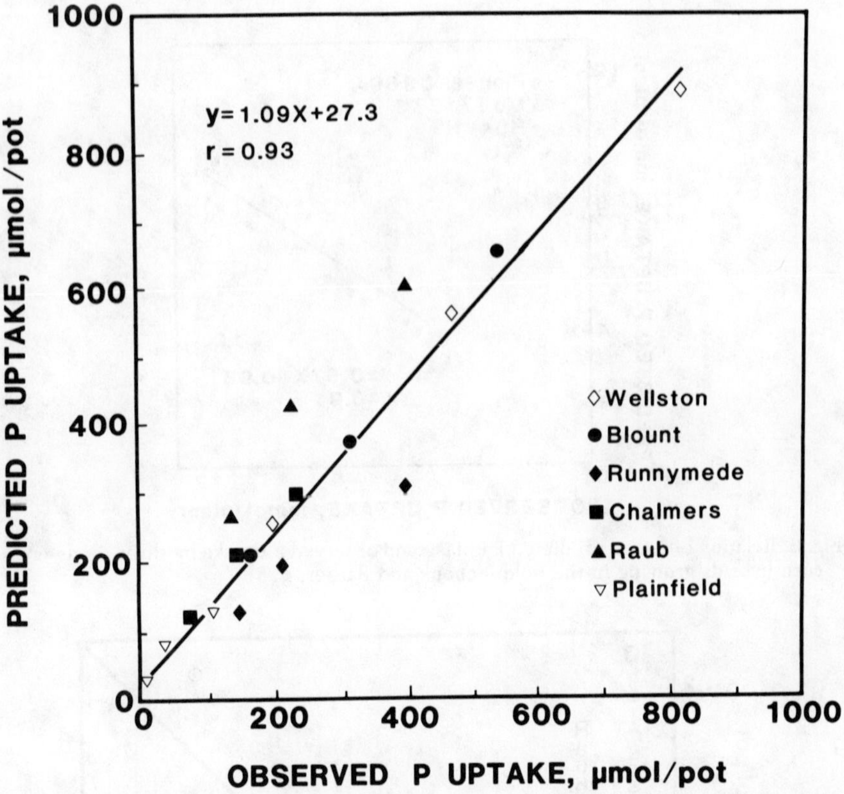

Fig. 2. Relation between predicted P uptake and observed P uptake by corn growing on six soils in 3-L pots in a growth chamber (Schenk and Barber, 1979).

adjacent roots. Silberbush and Barber (1983a) evaluated this model for predicting P and K uptake by three varieties of soybeans (*Glycine max* L. Merr.) grown to four ages in two soils, Raub silt loam (a fine-silty mixed, mesic Aquic Argiudoll) and Chalmers silt loam (a fine-silty mixed, mesic Typic Haplaquoll). The relation between predicted and observed P uptake shown in Fig. 4 indicates close agreement between observed and predicted uptake by soybeans grown in pot experiments in a controlled climate chamber. Similar results were obtained for K uptake by the same soil-plant system, however, in this case the agreement between predicted and observed uptake varied with soil. For Chalmers silt loam they agreed closely, while for Raub silt loam predicted uptake was an average of 37% larger than observed uptake. There was more variation in the data for K than those for P. Further investigations are needed to determine the reason for the variability in the correlations between predicted and observed K uptake.

Variations in root morphology are greater among species than within species. Itoh and Barber (1983a) investigated six plant species and used the Cushman model to relate predicted P uptake with observed P up-

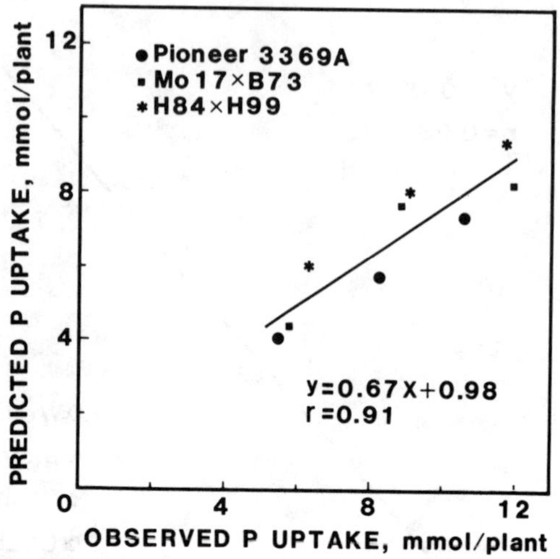

Fig. 3. Relation between predicted P uptake and observed P uptake by three single-cross corn hybrids growing in the field (Schenk and Barber, 1980).

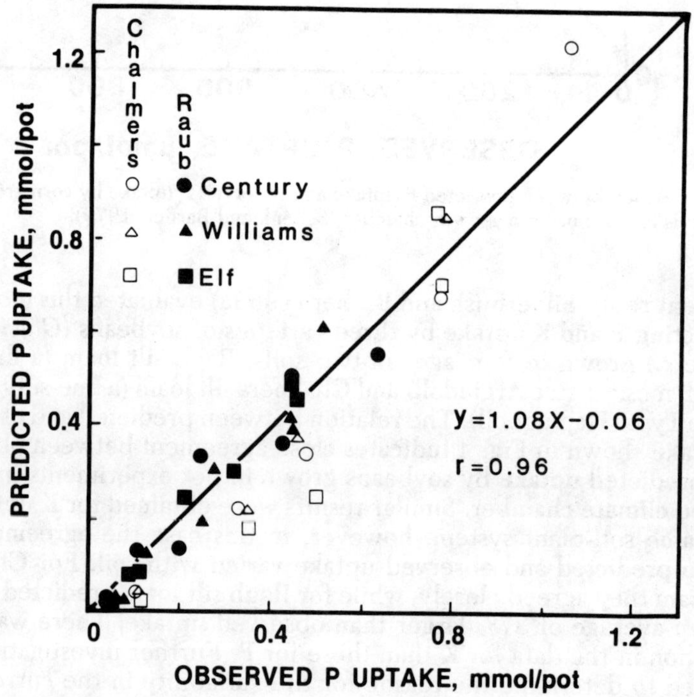

Fig. 4. Relation between predicted P uptake and observed P uptake by three soybean cultivars growing in two soils in 3-L pots in the growth chamber (Silberbush and Barber, 1983a).

Table 1. Morphological characteristics of the roots of six species grown for 30 to 50 days.

Plant	Root radius	Root weight / Total plant weight
	mm	g
Wheat (*Triticum aestivum* L.)	0.108 ± 0.003	0.34
Lettuce	0.124 ± 0.003	0.19
Russian thistle	0.056 ± 0.007	0.16
Tomato	0.107 ± 0.012	0.18
Onion (*Allium cepa* L.)	0.225 ± 0.017	0.16
Carrott (*Daucus carota* L.)	0.107 ± 0.018	0.24

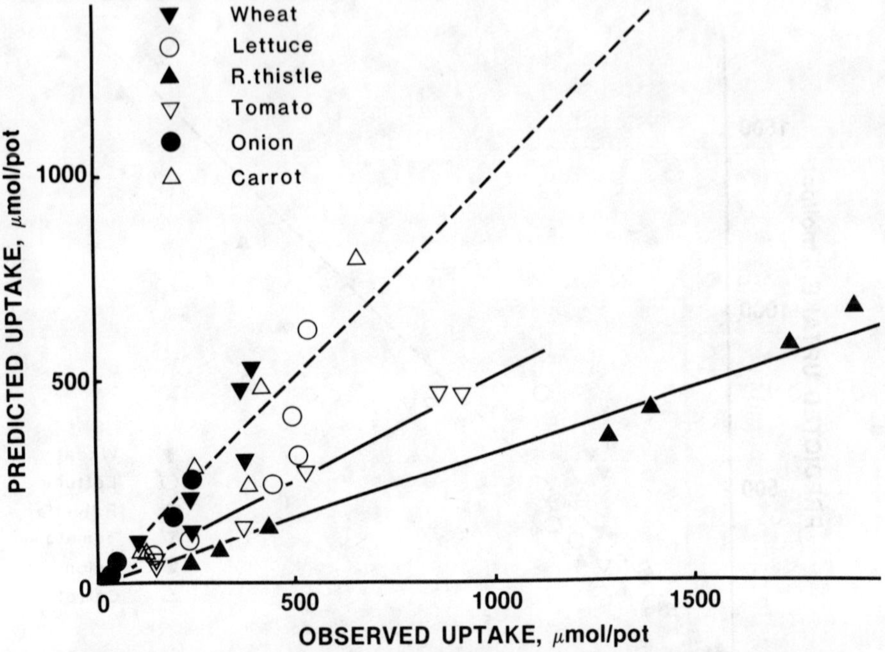

Fig. 5. Relation between predicted P uptake and observed P uptake by six plant species growing in Raub silt loam in 3-L pots in the growth chamber (Itoh and Barber, 1983).

take. The species used (Table 1) were selected because of their differential response to P added to the soil and their ability to grow on soils low in available P. For example, Russian thistle (*Salsola kali* L.) grew well on soils very low in available P, while lettuce (*Lactuca sativa* L.) gave a large growth response to added P when grown in soils low in available P. The relation between predicted and observed P uptake for the species used in this experiment is shown in Fig. 5. There were differences among species. Russian thistle and tomato (*Lycopersicon esculentum* Mill.) absorbed more P from the soil than predicted. The remaining four species gave observed P uptake that agreed with predicted P uptake. Itoh and Barber found that the roots of these species varied greatly in length and density of root hairs (Table 2). When the

Table 2. Characteristics of the root hairs of six species. Values are means and standard deviations (Itoh and Barber, 1983a).

Plant	No. of hairs	Length	Radius	Root hair surface / Root surface
	cm^{-1}	mm	mm × 10^3	
Wheat	560 ± 180	0.29 ± 0.12	5.7 ± 0.7	0.7
Lettuce	1270 ± 720	0.30 ± 0.12	4.8 ± 0.5	1.6
Russian thistle	890 ± 290	0.60 ± 0.15	3.9 ± 0.5	3.8
Tomato	1650 ± 420	0.43 ± 0.17	4.3 ± 0.1	2.5
Onion	1180 ± 110	0.04 ± 0.01	1.0 ± 1.4	0.2
Carrot	1810 ± 460	0.04 ± 0.02	4.0 ± 0.4	0.3

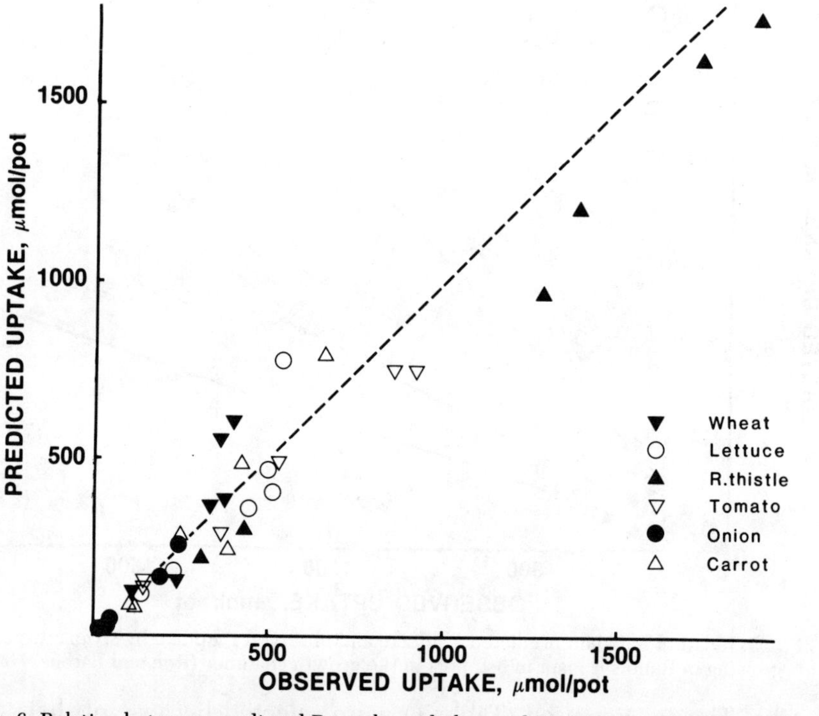

Fig. 6. Relation between predicted P uptake and observed P uptake by six plant species shown in Fig. 5 where root hair contribution to P uptake was included in predicted P uptake (Itoh and Barber, 1983).

effect of root hairs on uptake was incorporated into the model, predicted P uptake agreed with observed P uptake (Fig. 6). Hence, while it was not necessary to add root hair effects when calculating P and K uptake in some experiments, it was for other species and soils. The significance of root hairs will be shown later to depend on rate of diffusion of the ion, D_e, and length of the root hairs.

The mechanistic mathematical models of Claassen and Barber (1976) and Cushman (1979, case II) give predicted P and K uptake that agrees reasonably well with observed uptake. The effect of varying each of the parameters in the model can be determined with a sensitivity analysis. While it is not necessary to have a verified model to conduct sensitivity analyses, it makes the results more meaningful agronomically.

SENSITIVITY ANALYSES FOR K UPTAKE

Silberbush and Barber (1983b) conducted a sensitivity analysis of predicted K uptake by Williams soybeans grown on Raub silt loam in pots in a controlled climate chamber. The initial values used for each parameter were mean values taken from the experiment and are given in Table 3. The experimental results revealed linear root growth with time starting from 10 days after germination. So k represents the linear rate of root elongation with time between age of 10 and 20 days, and L_o is length of existing roots 10 days after germination. Each parameter was varied separately between 0.5 and 2.0 times the initial value and all other parameters were held at the level shown in Table 3 while predicted K uptake was calculated. The results of this sensitivity analysis are shown in Fig. 7.

Predicted K uptake was more sensitive to changes in root surface area, k, and r_o, than the same relative change in the other parameters. In the model when k is increased alone and r_1 remains constant, it assumes soil volume increases as root length increases, hence a linear increase in predicted K uptake is expected. Increasing r_o increases root surface area also in direct proportion to the increase in r_o. Since r_1 is held constant, soil volume decreases by the amount root volume increases as r_o increases. Also, changes occur in radial geometry of the root surface, hence the increase in predicted K uptake with increased r_o becomes curvilinear.

The major emphasis of this chapter is on root morphology, hence discussions will concentrate on the effects of k, r_o, and r_1 on predicted K uptake.

Sensitivity analysis can also be conducted where more than one parameter is changed. We can assume root length increases but soil volume remains constant. This is what is likely to occur in nature. Hence,

Table 3. Soil and plant K parameters for the Cushman mathematical model, their symbols and values for 'Williams' soybeans growing on Raub silt loam.

Symbol	Parameter	Initial value
D_e	effective diffusion coefficient	3.47×10^{-8} cm^2 s^{-1}
b	buffer power	24.0
C_{li}	initial conc. in soil solution	0.28 mmol L^{-1}
v_o	water influx	5.0×10^{-7} cm s^{-1}
r_o	root radius	0.015 cm
r_1	half the distance between root axes	0.2 cm
Imax	maximal influx	7.05×10^{-5} mmol m^{-2} s^{-1}
Cmin	minimal conc. where In = 0†	1.4×10^{-3} mmol L^{-1}
K_m	conc. − Cmin when In = 1/2 Imax	1.03×10^{-2} mmol L^{-1}
L_o	initial root length	250 cm
k	root growth rate	0.03 cm s^{-1}

† In = net influx at r_o.

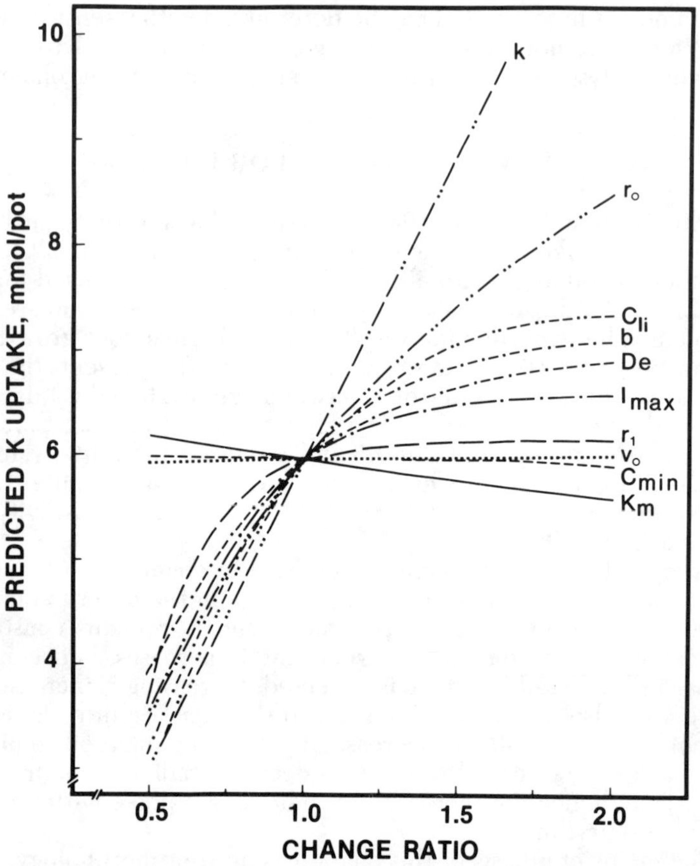

Fig. 7. Results of a sensitivity analysis for K uptake showing the relative effect of changing each model parameter separately from 0.5 to 2.0 of the initial value while all other parameters remain at the initial values shown in Table 1 (Silberbush and Barber, 1983b).

we decrease r_1 as k is increased in order to maintain root plus soil volume constant and we can also decrease r_o at the same time to keep root and soil volume constant. Figure 8 shows results of calculations using three sets of assumptions. Since r_1 is related inversely to k we can plot either parameter. We chose to use k. The curve labeled k is the same as in Fig. 7; only k was changed. The curve labeled k r_1 is where the root plus soil volume remained constant so as k increased, r_1 decreased. Since r_o was held constant, root volume in the pot increased and soil volume decreased. If we assume roots become thinner as the length is increased due to differing genetics, we can assume root volume remains constant. Curve kr_1r_o is where soil and root volume remain constant as root length, k, is changed. In this situation, the increase in predicted K uptake as k increased was less when soil plus root volume was held constant and the least where soil and root volume were each held constant. However, the differences were not large. The size of the difference depends on the degree of competition between

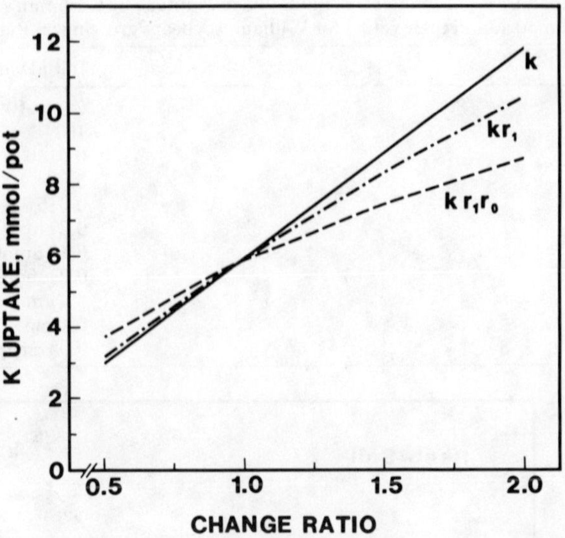

Fig. 8. Effect on predicted potassium uptake of varying k, rate of root growth, where: k, only k is varied; kr_1, r_1 is varied as k varies so that the sum of soil plus root volume remains constant; and kr_1 r_0, as k is varied, both r_1 and r_0 are varied so that soil and root volume each remain constant (Barber, 1984).

roots. While variation of root growth in a given species may not cover this range, it does show how differences in root morphology among species may affect the K they can recover from the soil. Grasses with longer finer roots usually compete for K uptake more favorably than thicker shorter rooted legumes.

SENSITIVITY ANALYSIS FOR P UPTAKE

A similar sensitivity analysis was conducted for predicted P uptake (Silberbush and Barber, 1983c). The initial values used for the sensitivity analyses were measured in an experiment investigating P uptake by soybeans growing in pots with 2.5 L of Raub silt loam in a controlled climate chamber. The initial values are given in Table 4. The values of D_e and C_{li} were much smaller than those for K while the value of b was much larger. Again as for K, changing the parameters k and r_o greatly affected predicted P uptake (Fig. 9). Uptake of P was more sensitive to soil supply parameters than to P influx kinetics of the soybean roots. The discussion of this sensitivity analysis will be confined to the effect of the root morphology parameters on predicted P uptake.

Predicted P uptake increased linearly with increases in k because the amount of soil which supplied P to the roots increased linearly. Predicted P uptake also increased rapidly as r_o increased because root

Table 4. Plant and soil parameters for simulation of P uptake by Cushman's mechanistic mathematical model, and their values for Williams soybeans growing on Raub silt loam.

Symbol	Initial value
D_e	2.3×10^{-9} cm^2 s^{-1}
b	163
C_{li}	0.0136 mmol L^{-1}
v_o	5.0×10^{-7} cm s^{-1}
r_o	0.015 cm
r_1	0.2 cm
Imax	6.43 nmol m^{-2} s^{-1}
Cmin	0.2 μmol L^{-1}
Km	5.4 μmol L^{-1}
L_o	250 cm
k	0.03 cm s^{-1}

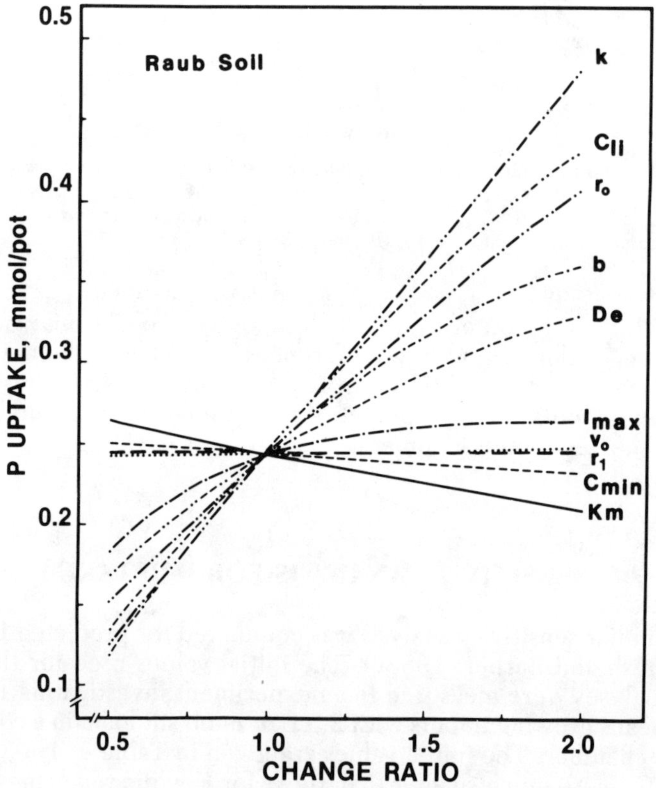

Fig. 9. Results of a sensitivity analysis for P uptake showing the relative effect of changing each model parameter separately from 0.5 to 2.0 of the initial value while leaving all other parameters at the initial levels shown in Table 2 (Silberbush and Barber, 1983c).

surface area increased linearly with increases in r_o. In the usual situation where plants are growing in the field, soil volume will remain constant so as k increases, r_1 decreases according to the relation $r_1 \sim 1/k^{1/2}$. Calculations were also made where root volume was held constant so r_o was varied according to $r_o \sim 1/k^{1/2}$. Figure 10 gives the sensitivity

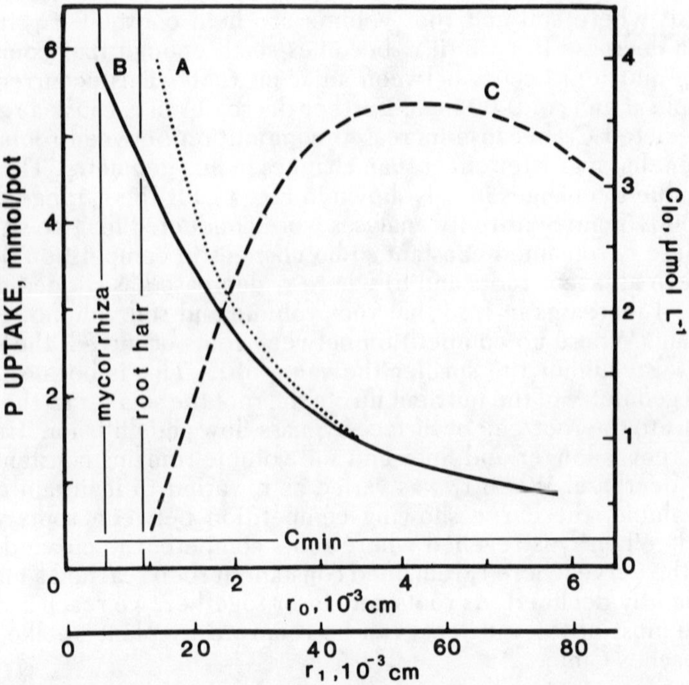

Fig. 10. Effect of changing r_o on calculated P uptake where root volume growth was constant (k increases as r_o decreases) where, A) r_1 was held constant (soil volume increased as k increased), B) r_1 decreased as K increased so soil volume was constant, and where curve C is C_{lo}, P concentration at the root surface after 10 days, under conditions of curve B (Silberbush and Barber, 1983c).

results for P uptake as affected by r_o where curve A, root volume, was held constant but soil volume varied; and curve B, root and soil volume were held constant as r_o varied from 0.006 cm, the radius of grass roots, to 0.0003 cm, the radius of mycorrhizal hypha. Decreasing r_o increases k and increases root surface area for absorption. This increased predicted P uptake as roots became longer and finer but had the same volume. Values for r_1 only had an effect on predicted P uptake when r_1 fell below 0.05 cm for the values of D_e and t used in this research. Mean distance for ion diffusion is $(2D_et)^{1/2}$. The value used for D_e was 2.3×10^{-9} cm^2 s^{-1} and for t was 10 days, hence mean diffusion distance was 0.063 cm. This is the reason r_1 only became a factor affecting predicted P uptake measurably when it fell below 0.05 cm. In addition to increasing root surface area, roots with a smaller r_o will have a higher C_{lo} nutrient concentration at the root surface, because of the effect of radial geometry on nutrient flux to the root. The C_{lo} values, which had reached an approximate constant level after 10 days, are shown in relation to r_o by curve C of Fig. 10. Curve C indicates that for the parameters used, C_{lo} reached a maximum when r_o was 0.004 cm. The initial r_o value shown in Table 4 was 0.015 cm, hence Fig. 10 shows the sensitivity analysis for the lower range of r_o values. In the sensitivity analysis where soil and root volume are held constant, C_{lo} increases with a decrease in r_1 until r_1 becomes small enough that competition for the nutrient occurs between adjacent roots. This occurred in this example at an r_o of 0.004 cm. Further decreases in r_o cause a reduction in predicted C_{lo} because increased competition between roots for the nutrient has a greater effect than changes in flux geometry. The change in C_{lo} due to changes in r_o is shown in Fig. 11 for the r_o range of 0.015 to 0.00015 cm. Sensitivity analyses were conducted for two situations: a) where r_1 remained constant so no changes in competition occurred between adjacent roots and b) where r_1 decreased as r_o decreased because of increases in k so that root volume and soil volume remained constant. Where no competition between roots occurred, the value of C_{lo}/C_{li} was higher the smaller the value of r_o. This is because for the radial geometry of the nutrient flux to the root the smaller r_o, the greater the flux to the root/cm² of surface by mass-flow and diffusion. However, if the root is longer and finer and soil volume remains constant values for r_1 decrease. When r_1 was varied as r_o varied to maintain constant soil volume, the curve showing competition between roots was obtained. When r_1 is reached where roots compete, the curve departed from the curve where r_1 remained constant, it then reached a maximum and rapidly declined. As roots get closer together, we reach a situation where most of the nutrients can be removed by plant uptake and C_{lo} approaches Cmin.

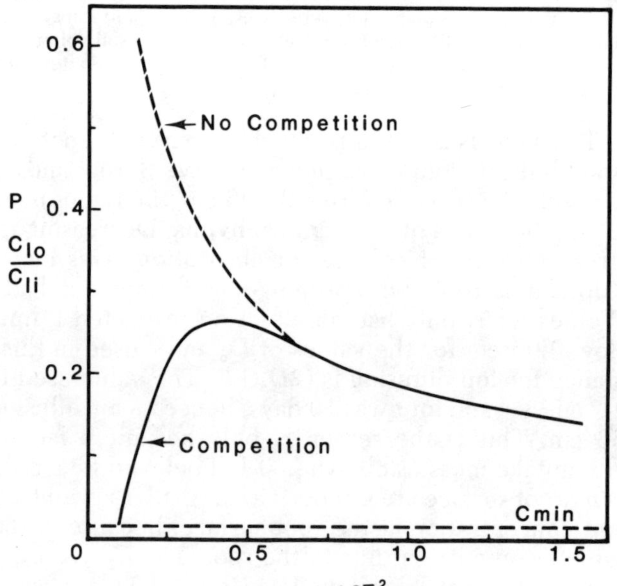

Fig. 11. Calculated C_{lo}/C_{li} where r_o ranged from 0.00015 to 0.015 cm, k increased so that root volume remained constant, when r_1 was held constant and no root competition occurred; and when r_1 changed so that soil volume remained constant and root competition occurred.

SENSITIVITY ANALYSIS OF N UPTAKE

The parameters describing NO_3^- uptake by a mathematical model vary greatly from those for P or K, hence they are useful for illustrating the effect due to large differences between nutrients in parameters size. A sensitivity analysis for NO_3^- was conducted using the initial parameters shown in Table 5. The parameters were measured for corn growing in Raub silt loam. The results of varying each parameter separately from 0.5 to 2.0 of the initial value while holding the remaining values at the initial level are shown in Fig. 12.

The sensitivity analysis indicated predicted NO_3^- uptake was very sensitive to values of k, Imax, and r_o, but not to changes in the size of the remaining variables in the range investigated. The sensitivity of k was because r_1 was held constant and soil volume increased with k, hence amount of nitrate present also increased. If soil volume is held constant, increasing k has only a small effect on uptake because D_e is large and the NO_3^- can readily reach the root. Increasing Imax increases uptake because the NO_3^- level at the root is usually high enough for net influx, I_n, to be near Imax. This is because Km is small relative to C_{li}. This is also why varying D_e, v_o, Km, and Cmin had so little effect when varied in the 0.5 to 2.0 range. Increasing r_o increased uptake because of the increase in root surface. However, this effect is also dependent upon the plant being able to utilize this increased influx and maintaining Imax.

COMPARISON OF THE SENSITIVITY ANALYSES FOR NO_3, K, AND P

The average values of the model parameters for NO_3, K, and P vary widely among these nutrients because of the large differences in strengths of adsorption for these nutrients by the soil as well as differences in the quantities needed by the plant. Comparisons were made where one plant species was used so that root morphology and water

Table 5. Plant and soil parameters used for simulation of NO_3^- uptake by Cushman's model. Values were obtained for corn growing in Raub silt loam.

Parameter	Initial value
D_e	2.5×10^{-6} cm^2 s^{-1}
b	1.0
C_{li}	5 mmol L^{-1}
v_o	1×10^{-7} cm s^{-1}
r_o	0.02 cm
r_1	0.3 cm
Imax	100 nmol m^{-2} s^{-1}
Cmin	0.002 μmol cm^{-3}
Km	0.025 μmol cm^{-3}
L_o	20 cm
k	0.03 cm s^{-1}
t	8.64×10^5 s

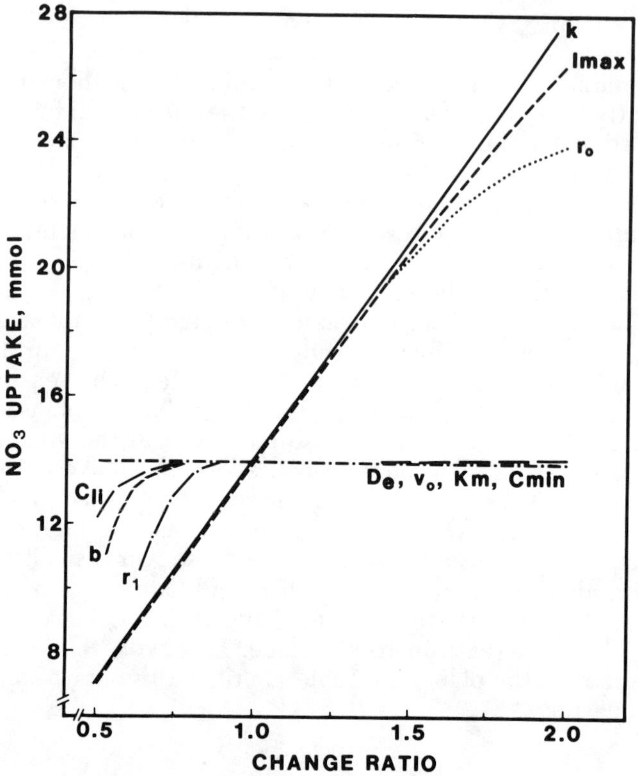

Fig. 12. Sensitivity analysis of model parameters used to predict NO_3^- uptake by corn grown in Raub silt loam where each parameter was varied from 0.5 to 2.0 of initial level while remaining parameters remained at initial levels shown in Table 5.

Table 6. Comparison of plant nutrient influx and soil supply parameters for NO_3^-, K, and P. Values were measured in experiments with Raub silt loam.†

Parameter and units	Nutrient		
	NO_3	K	P
D_e cm^2 s^{-1} × 10^8	250	3.47	0.23
b	1	24	163
C_{li} mmol L^{-1}	5	0.28	0.0136
Imax nmol m^2 s^{-1}	100	70.5	6.4
Cmin μmol L^{-1}	2	1.4	0.2
Km μmol L^{-1}	25	10	5

† Values in common were: v_o, 1 × 10^{-7} cm s^{-1}; r_o, 0.015 cm; r_1, 0.25 cm; L_o, 250 cm; k, 0.03 cm s^{-1}; t, 432 × 10^5 s.

influx would be common for the sensitivity analysis of all three nutrients. The parameters used are given in Table 6.

A comparison was made of the effect of reducing r_o on C_{lo}/C_{li} at the root surface where soil and root volume remained constant (Fig.

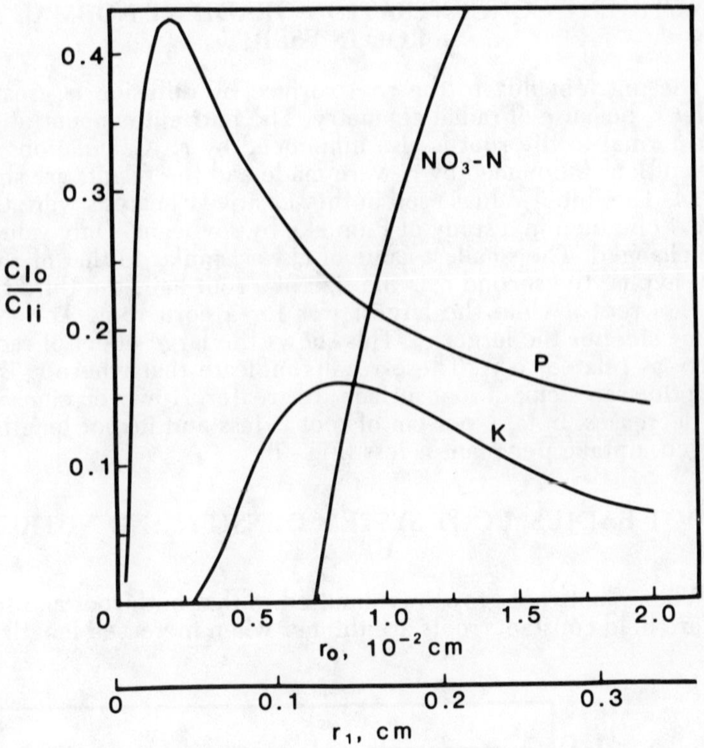

Fig. 13. Comparison of the C_{lo}/C_{li} for NO_3^-, K, and P where r_o and r_1 are changed to keep soil and root volume constant.

13). The C_{lo}/C_{li} value is after 5 days of uptake by this particular section of root with combinations of r_o and r_1 values shown.

Values of C_{lo}/C_{li} increase as r_o decreased until r_1 became small enough that root competition occurred in 5 days. This depended on the mean diffusion distance. Using the values of D_e and t in Table 6, we obtained mean diffusion distances of 1.47 cm for NO_3^-, 0.17 cm for K, and 0.045 cm for P. The increasing portion of the curve for NO_3 is not shown because it was at values of r_1 greater than the initial value. As r_1 is decreased, a maximum value of C_{lo}/C_{li} is reached and then competition between roots reduces it. This value of r_1 corresponds with the mean diffusion distances for NO_3^-, K, and P that we calculated. As r_1 values decrease further competition occurs and C_{lo}/C_{li} decreases rapidly until C_{lo} approaches Cmin. The value of r_1 where C_{lo}/C_{li} starts to decrease becomes smaller with a decrease in D_e for the nutrient in the soil.

Of the three nutrients, reduction in r_o for P has the greatest effect on C_{lo}/C_{li} because with its low D_e a smaller value of r_o can be reached before competition among adjacent roots begins as r_1 decreases and C_{lo}/C_{li} decreases.

NUTRIENT CONCENTRATION GRADIENT NORMAL TO ROOTS VS. R_o

The nutrient flux to the root surface by diffusion is greater the smaller r_o because of radial geometry. The nutrient concentration gradient normal to the root is also influenced by r_o. Calculations of gradients of P as influenced by r_o were made and the results are shown in Fig. 14. The initial values used in the sensitivity analysis were those in Table 4 obtained in a study of P uptake by soybeans. Only values of r_o were changed. The smallest value of r_o was similar to that of a mycorrhizal hypha, the second r_o is the size of a root hair, the third the size for grass roots, while the largest was for a corn root. The gradient became steeper the larger r_o. This shows the large effect of radial geometry as related to r_o. These results indicate that where r_o is small, absorption rates/cm² of root surface are greater. However, since surface area decreases, uptake per cm of root is less and if root length is not changed, uptake per plant is less (Fig. 9).

ROOT RADIUS, ROOT SYSTEM DENSITY, AND NUTRIENT UPTAKE

When plant root growth is simulted so that both root and soil volume are held constant, roots are thinner when increased length is sim-

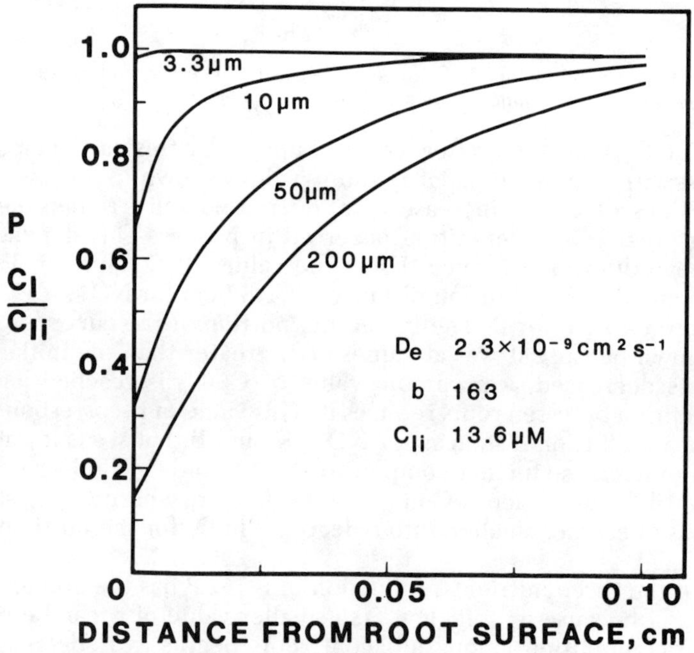

Fig. 14. Effect of root radius on relative depletion of C_l of P extending radially from the root surface after 10 days uptake. Soil and plant parameters other than root radius were held constant and are given in Table 4.

ulated. The increased length increases root system density so that r_1, the half-distance between root axes is reduced. As r_1 is decreased, there is greater competition between roots for nutrients in the soils.

Calculations were made of the effect of changing root radius and density on P, K, and NO_3^- uptake. Calculations were for 5 days where root and soil volume remained constant. The results are shown in Fig. 15. In order to compare nutrients, uptake was calculated as a fraction of the total diffusible nutrient in the soil. Practically all of the NO_3^- was predicted to be absorbed in 5 days by relatively thick and short root systems. However, the calculated absorption was only a small portion of the diffusible P. A very thin and dense root system with an r_o of the size of root hairs or mycorrhiza hyphae would be required to take up a high proportion of the P in the soil in 5 days. The bars in Fig. 15 represent reported radii ranges for roots of different species and for root hairs.

Differences in the position of the lines among the three nutrients is mainly due to differences in their D_e and b values in the soil. Hence,

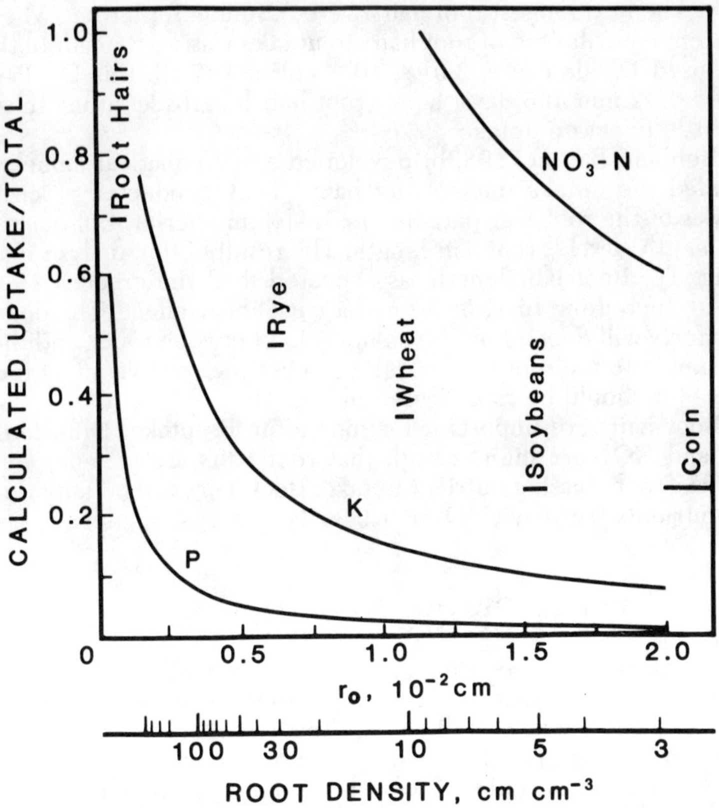

Fig. 15. Comparison of effect of changing r_o on calculated uptake of NO_3^-, K, and P where root and soil volume are kept constant. Bars represent range of root r_o for species shown and root hairs.

effect of r_o and length of root system on nutrient uptake varies with the nutrient considered as well as with plant species.

ROOT HAIR SENSITIVITY ANALYSIS

The foregoing discussion assumed roots were smooth uniform cylinders of average r_o that absorbed nutrients uniformly over their surface. Many roots, however, have root hairs that extend out from the root surfaces. Root hairs will influence uptake of some nutrients. Phosphorus uptake is often increased by root hairs. This is because D_e for P is small. Whether root hairs affect uptake or not appears to depend primarily on the relation between root hair length and D_e. If nutrients diffuse greater distances than the length of the root hair, then root hairs do not contribute as much to uptake. Itoh and Barber (1983a) measured uptake by six species that varied in root hair length and density. Two species had root hairs that increased P uptake. They were Russian thistle and tomato. Their root hair lengths were 0.60 and 0.43 mm, respectively. The next longest root hairs were 0.30 mm for lettuce where the apparent contribution of root hairs to uptake was very small. In the soil they used D_e for P was 2.0×10^{-9} cm^2 s^{-1}. With this D_e, P would diffuse 0.42 mm in 5 days, hence root hair lengths less than this value had little effect on uptake.

Itoh and Barber (1983b) developed a mathematical model which included the uptake due to root hairs. They conducted a sensitivity analysis of the root hair parameters, Imax, number of root hairs/cm of root length, and L_h, root hair length. The result of this analysis is shown in Fig. 16. Root hair length, as expected, had the greatest effect on uptake. Increasing Imax or N_h had a curvilinear effect. The degree of the effects will depend on the competition between roots, and the rate of P supply to the root by the soil, however the relative effect for each parameter should be as indicated in Fig. 16.

Root hairs are important for increasing P uptake. Diffusion rates of K and NO_3^- are high enough that root hairs are of much less importance in increasing nutrient uptake. Root hairs will be important for any nutrients that have a D_e of 1.0×10^{-9} cm^2 s^{-1} or less.

ROOT MORPHOLOGY AND CROP YIELD

Varieties of species vary in yield. Root growth is one of the parameters that can affect nutrient supply to the plant and plant growth. It is one of several factors that may affect yield. Silberbush and Barber investigated root growth of soybean varieties and found differences between soybean varieties. They related the soybean yield to total root length at the R6 stage of growth. The results are shown in Fig. 17. There was clearly a relation between greater root growth and higher yield.

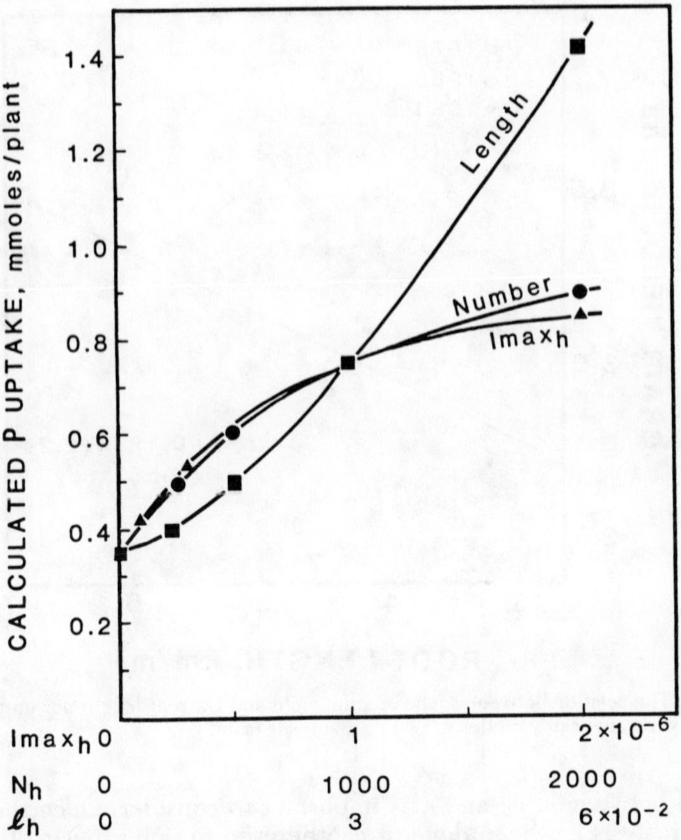

Fig. 16. Effect of varying Imax for root hairs, root hair no./cm, and root hair length on calculated P uptake from soil (Itoh and Barber, 1983b).

The relation between root length and crop yield indicates that root growth is significant in determining nutrient supply to the shoot and crop yield.

CONCLUSIONS

Mechanistic mathematical models are useful research tools for evaluating the parameters that may be most significant in affecting growth of the plant. In the evaluation of root parameters in relation to nutrient uptake, each was varied by the same relative amount. In nature, it may be easier to vary one parameter than another. For example, addition of fertilizer will increase C_{li} and this can be easily done. Varying root length using a breeding program is more difficult. However, it is important to determine which parameters are the most sensitive and then to investigate the range of variation that may be possible for that particular parameter. The interdependence of parameters such as was dis-

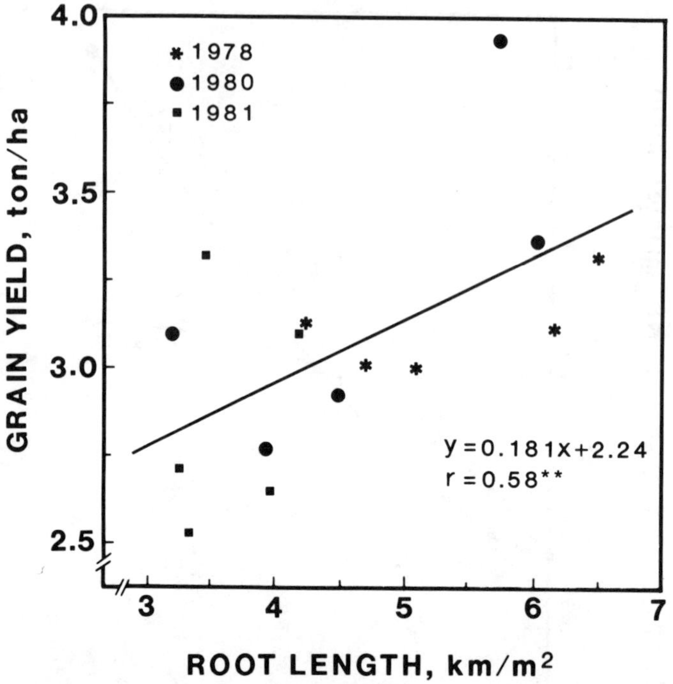

Fig. 17. The relation between soybean grain yield and L_A, root length per unit soil area for several soybean varieties grown on Raub silt loam.

cussed between k, r_1, and r_o is important to consider. Once the sensitivity analysis has been done, it can be used to point toward the lines of research that will have the most promise of success.

REFERENCES

1. Barber, S.A. 1984. Soil nutrient bioavailability: A mechanistic approach. Wiley-Interscience, New York, NY.
2. ----, and J.H. Cushman. 1981. Nitrogen uptake model for agronomic crops. p. 382–409. *In* I.K. Iskandar (ed.) Modeling wastewater renovation-land treatment. Wiley-Interscience, New York.
3. Claassen, N., and S.A. Barber. 1976. Simulation model for nutrient uptake from soil by a growing plant root system. Agron. J. 68:961–964.
4. Cushman, J.H. 1979. An analytical solution to solute transport near root surfaces for low initial concentration. I. Equations development. Soil Sci. Soc. Am. J. 43:1087–1090.
5. Itoh, S., and S.A. Barber. 1983a. Predicted and observed uptake by six plant species as related to root hairs. Agron. J. 75:457–461.
6. ----, and ----. 1983b. A numerical solution of whole plant nutrient uptake for soil-root systems with root hairs. Plant Soil 70:403–413.
7. Nye, P.H., J.L. Brewster, and K.K.S. Bhat. 1975. The possibility of predicting solute uptake and plant growth response from independently measured soil and plant characteristics I. Theoretical basis of the experiments. Plant Soil 42:161–170.

8. Schenk, M.K., and S.A. Barber. 1979. Root characteristics of corn genotypes as related to phosphorus uptake. Agron. J. 71:921–924.
9. ----, and ----. 1980. Potassium and phosphorus uptake by corn genotypes grown in the field as influenced by root characteristics. Plant Soil 54:65–76.
10. Silberbush, M., and S.A. Barber. 1983a. Prediction of phosphorus and potassium uptake by soybeans with a mechanistic mathematical model. Soil Sci. Soc. Am. J. 47:262–265.
11. ----, and ----. 1983b. Sensitivity analysis of parameters used in simulating potassium uptake with a mechanistic mathematical model. Agron. J. 75:851–854.
12. ----, and ----. 1983c. Sensitivity of simulated phosphorus uptake to parameters used by a mechanistic-mathematical model. Plant Soil 74:93–100.

Chapter 5

pH Changes and Phosphate Solubilization near Roots—An Example of Coupled Diffusion Processes[1]

P.H. NYE[2]

It is now well established that the concentration profiles of important nutrient ions, such as K^+ and NO_3^-, near absorbing roots, and their rates of uptake can be satisfactorily explained by a model in which the ion moves to the root by mass flow and diffusion. The root serves as a sink whose uptake rate is a function of the concentration of the ion in solution at its surface (Nye and Tinker, 1977; Barber and Silberbush, 1984). The model assumes that local equilibrium, characterized by a unique sorption isotherm, is rapidly established between soil solution and solid surfaces. However, the sorption isotherm depends on concentrations of other ions in the system, and these may be changed in the rhizosphere soil by root uptake or excretion. The concentration of Ca^{++} and Mg^{++} in the soil solution is usually above $10^{-3}M$, and is high enough to be changed little by root uptake. But, the solution concentration of sparingly soluble ions, such as phosphate and trace

[1] Contribution of the Dep. of Agricultural and Forest Sciences, Univ. of Oxford, England.

[2] Reader in soil science, Dep. of Agricultural and Forest Sciences, Univ. of Oxford, England.

Copyright © 1984 Soil Science Society of America, Crop Science Society of America, and American Society of Agronomy, 677 South Segoe Road, Madison, WI 53711. *Roots, Nutrient and Water Influx, and Plant Growth.*

elements, may depend sensitively on pH and complexing ions. It is well established that roots can change the rhizosphere pH by as much as 1 to 2 units (Nye, 1981) and there has been much speculation about root exudation of solubilizing complexes. This chapter considers whether these additional complications can be incorporated into the quantitative model that has proved so satisfactory for simple nutrient ions.

pH CHANGES ACROSS THE RHIZOSPHERE

If plants absorb N in the form of NO_3^-, in general they absorb more anions than cations and excrete HCO_3^- to maintain electrical neutrality across the root-soil interface. If their source of N is NH_4^+, or N_2 fixed symbiotically, they absorb more cations than anions and excrete H^+. The amounts that may be involved are large: Cunningham (1964) found 62 common crop and arable weed species contained on average 3.6 meq of absorbed anions ($NO_3^- + SO_4^{2-} + H_2PO_4^- + Cl^-$) and 2.5 meq of absorbed cations ($K^+ + Na^+ + Ca^{2+} + Mg^{2+}$) per gram dry tops, implying that 1.1 meq of HCO_3^- are excreted per gram, or nearly one HCO_3^- ion for every two cations absorbed. For plants fixing N symbiotically the available evidence shows that about 1 meq of H^+ is exuded per gram of plant dry weight (Nyatsanga and Pierre, 1973; Israel and Jackson, 1978). The consequent pH changes in the soil around the roots have been detected experimentally (Riley and Barber, 1969; Smiley, 1974; Bagshaw et al., 1972).

PREDICTION OF THE pH CHANGE

The course of development and extent of the pH change around roots may be predicted from a theoretical treatment of acid-base diffusion in soil (Nye, 1981). This theory uses a soil acidity diffusion coefficient, D_{HS}, in the diffusion equation

$$\partial pH/\partial t = (\partial/\partial x)[D_{HS}(dpH/dx)] . \qquad [1]$$

It has been shown that for practical purposes (Nye, 1972)

$$D_{HS} = (2.303\theta f/b_{HS})\,(D_{LH}[H_3O^+] + D_{LC}[HCO_3^-]) , \qquad [2]$$

where θ = soil water fraction by volume,
f = diffusion impedance factor,
D_{LH}, D_{LC} = the diffusion coefficients of H_3O^+, HCO_3^- in free solution,
b_{HS} = soil pH buffer power (− d[HS]/d pH), and
HS = soil acid.

The explanation of Eq. [2] is that in an acid-base diffusion system a small portion of soil (S) may gain acidity either by access of hydronium ions ($S + H_3O^+ = SH^+ + H_2O$), or by dissociation of carbonic acid (derived from CO_2) and removal of bicarbonate ions through the soil solution ($S+H_2CO_3=SH^++HCO_3^-$). The relative importance of the two dominant mobile acid-base carriers in the soil solution, H_3O^+-H_2O and H_2CO_3-HCO_3^-, depends on their diffusion coefficients in solution and their concentrations, as expressed in the terms in the bracket in Eq. [2] and plotted in Fig. 1. It will be seen that for a given pressure of CO_2 there is a pH in the range 5 to 6 at which both H_3O^+ and HCO_3^- are at a low concentration in the soil solution, and the soil acidity diffusion coefficient is at a minimum. Its value increases rapidly as the pH changes from that at the minimum. This implies that if there is a flux of acid or base through a soil the pH gradients will be steep in the neighborhood of pH 5 to 6 and gentle at low or high pH. A pH 5 to 6 is a particularly interesting plant physiological range, which, by this argument, is readily altered at the root-soil interface by release of H^+ or HCO_3^-. The effect of release of H^+ at the root surface on pH profiles in the rhizosphere is illustrated in Fig. 2.

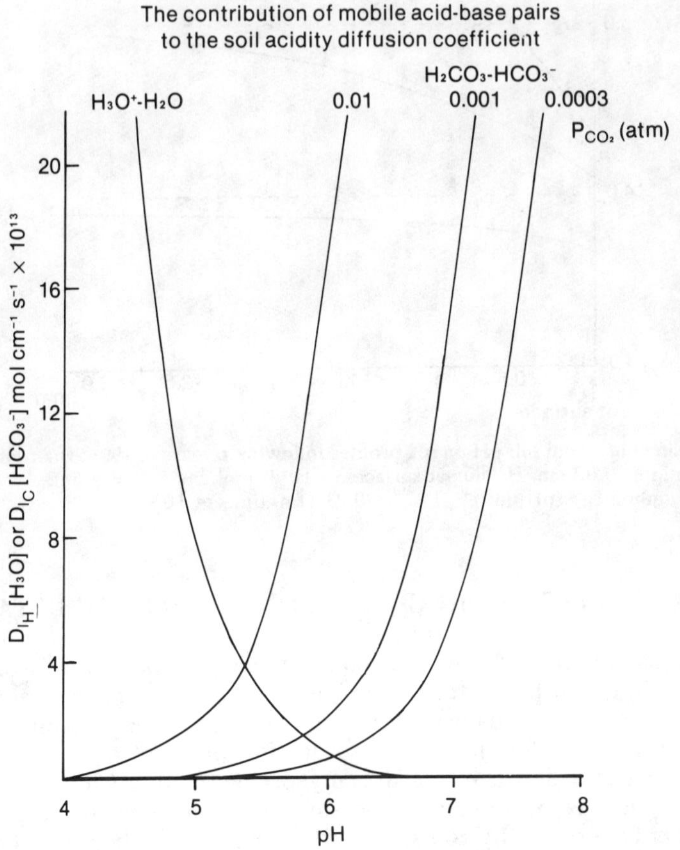

Fig. 1. Relative importance of H_3O^+ or HCO_3^- in acid-base transport at different pH (see Eq. [2]).

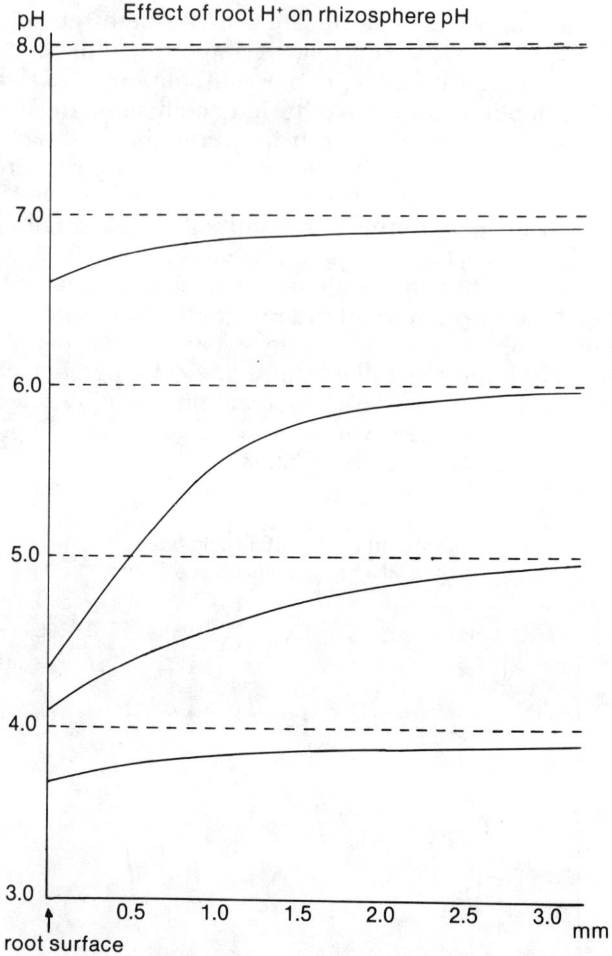

Fig. 2. Effect of initial soil pH on pH profiles following release of H^+ at the root surface. (root radius 0.02 cm, H^+ flux at surface 3×10^{-12} mol $cm^{-2}s^{-1}$, $\theta = 0.2$, f = 0.1, b_{HS} = 11 μmol mL^{-1} (pH unit)$^{-1}$, P_{CO_2} = 0.01 atm. time = 10^6s.)

THE EFFECT OF pH CHANGE ON PHOSPHATE UPTAKE

It has long been known that some species are particularly effective in extracting phosphate from P deficient soil, and there has been much discussion of the mechanism involved. Among the adaptations or mechanisms suggested for increasing absorption per unit of root are: finer roots, long and dense root hairs, mycorrhizas, high Ca uptake, solubilizing chelates produced by roots or rhizosphere bacteria, phosphatases, acid excretion by roots, or acid production by bacteria. Of these possibilities, it is now well established that mycorrhizas and fine roots and root hairs can promote P absorption, but other mechanisms are less firmly established. Our attempts to predict the rate of P uptake by rape (*Brassica napus* L.) and onions (*Allium cepa* L.) using independently determined soil and plant parameters in a model similar to that described by Barber in this symposium agreed reasonably well with experiment at moderate P levels, but greatly underpredicted the uptake from the soil at a low P level (Brewster et al., 1975, 1976). Further, rape was very much more efficient in absorbing phosphate than onion per unit length of root (Table 1) though its roots are not mycorrhizal. Subsequent experiments on rape were designed to achieve a high rooting density, and sufficient rhizosphere soil for detailed study, by growing the seedlings in a thin layer of soil between plexiglass plates. The soil, pH = 6.0, was supplied with a complete nutrient solution (lacking phosphate), and the only source of N was NO_3^-. It was found (Grinsted et al., 1982a, 1982b, and 1982c; Hedley et al., 1983) that:

(a) in the soil amended with phosphate the pH remained steady, in the unamended soil it fell (Fig. 3),
(b) the change in pH could be accounted for by the excess of cations over anions absorbed by the rape with consequent excretion of H^+, and the measured pH buffer capacity of the soil (Fig. 4),
(c) in the illitic sandy loam soil (brown earth) used lowering of pH markedly increased the concentration of phosphate in the soil solution; this effect was sufficient to account for the phosphate absorbed by the rape in excess of that predicted (Fig. 5),
(d) there was no appreciable increase in the very small amounts of organic acids or their anions that were extracted from the rhizosphere soil (Table 2). The total uronic acids in the rhizosphere soil differed little from the control soil, and
(e) the acid phosphatase activity in the unamended soil increased. However, this appeared to be a consequence of the low P concentration, since the level of organic P in the rhizosphere soil actually increased slightly, probably because of root hair debris and microbial immobilization of soil inorganic P.

In such instances as this, our existing model of solute diffusion to a root must clearly be modified to take account of the simultaneous release of a solubilizing agent, such as H^+, from the root surface.

Table 1. Phosphorus uptake by seedlings from Begbroke sandy loam.

Exchangeable P in soil	2.4 μmol/mL	
Soil solution P	2.6 μmol/L	
	Onion	Rape
Growth period	12 day	5 day
Length of primary roots at end of growth period	13.5 cm	14.5 cm
Length of laterals	0	12 cm
Phosphorus uptake by plant (mol × 10^7)		
Shoot	0.17	9.7
Root	0.08	0.8

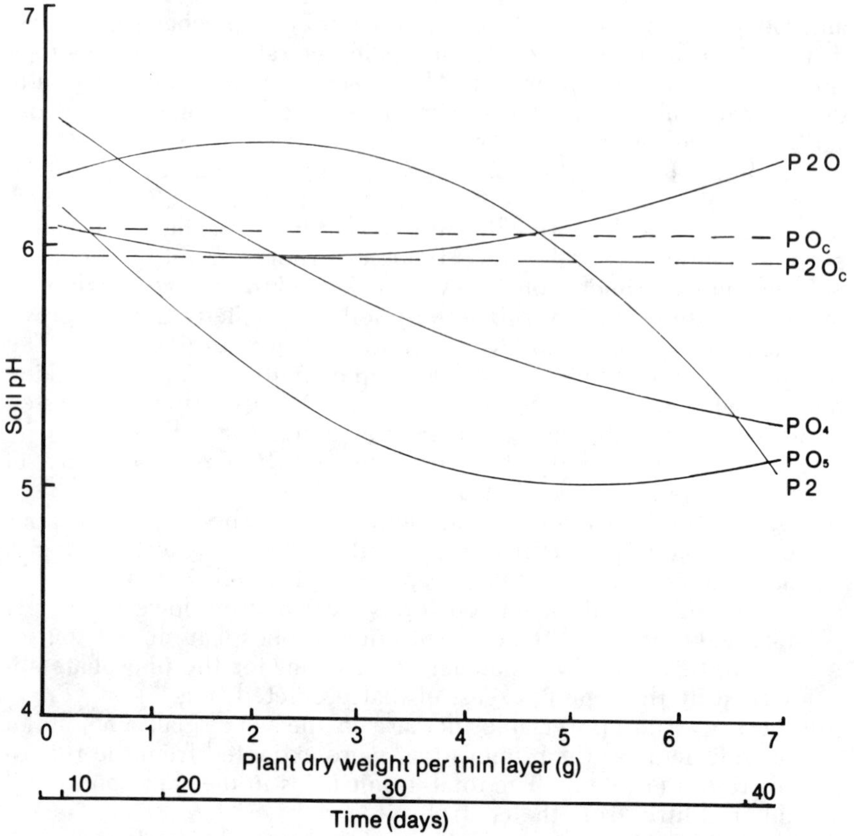

Fig. 3. The rhizosphere soil pH at a given plant dry wt. PO_4, PO_5, unamended soil; P2, 2 μmol P/g soil; P20, 20 μmol P/g soil; PO_c, $P20_c$, unplanted controls.

A MODEL FOR THE COUPLED DIFFUSION OF SOIL PHOSPHATE AND SOIL ACID

The differential equations for the diffusion of two solutes which interact with each other in the soil may be derived by the methods used by Henry (1939) to solve an analogous problem of the simultaneous transport of heat and moisture in cotton (*Gossypium hirsutum* L.) bales. The theory leads to two simultaneous second order differential equations

$$D_P(1-\upsilon\lambda)[(\partial^2 pP)/\partial x^2] - (\partial/\partial t)(pP-\lambda pH) = 0 \text{ and} \quad [3]$$

$$D_{HS}(1-\upsilon\lambda)[(\partial^2 pH)/\partial x^2] - (\partial/\partial t)(pH-\upsilon pP) = 0\ ; \quad [4]$$

where pP, pH = negative logarithm of the concentration of phosphate ions and hydronium ions in the soil solution,

λ, υ = interaction coefficients $(\partial pP/\partial pH)_P$, $(\partial pH/\partial pP)_{HS}$, and

D_P, D_{HS} = diffusion coefficients of P and HS in the soil.

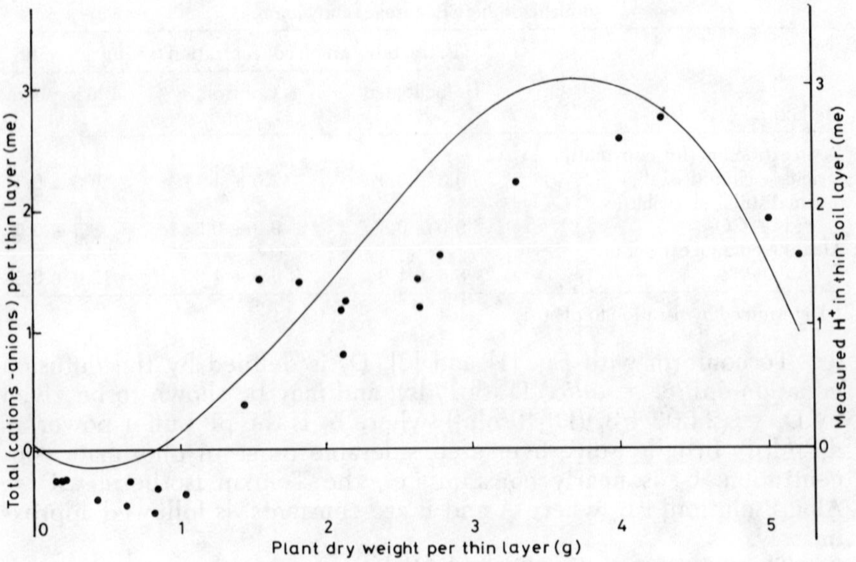

Fig. 4. Comparison of the difference between plant cation and anion uptake per thin layer, and the amounts of H^+ and OH^- required to produce the soil pH change, as calculated from the soil pH buffer capacity. ——Total (cations-anions) per thin layer. • Total amount of H^+ or OH^- required.

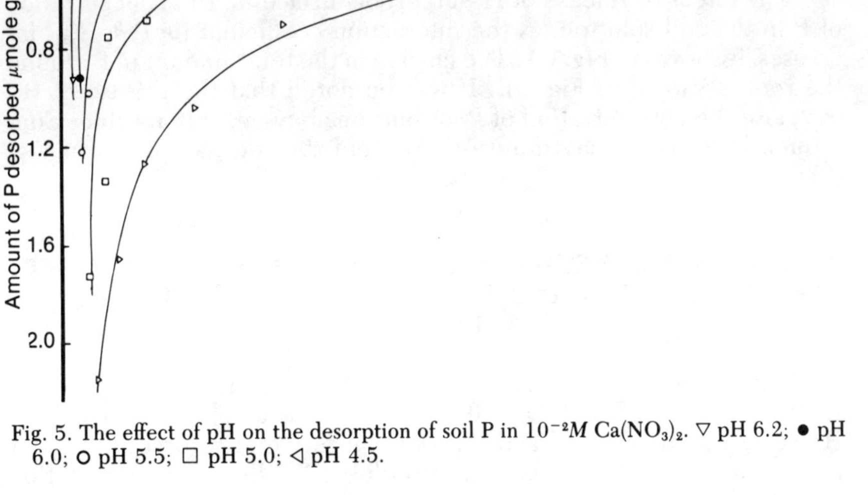

Fig. 5. The effect of pH on the desorption of soil P in $10^{-2}M$ $Ca(NO_3)_2$. ▽ pH 6.2; ● pH 6.0; ○ pH 5.5; □ pH 5.0; ◁ pH 4.5.

Table 2. Amounts of organic acids and anions extracted from rhizosphere and nonrhizosphere Begbroke sandy loam.

Method	Extractable anion concentration (μequiv g^{-1} soil)		
	Unincubated soil	Control soil	Rhizosphere soil
Resin extraction (for nonvolatile acids excluding oxalic)	9.3 ± 0.9†	12.0 ± 1.1†	9.3 ± 0.5†
Steam distillation (volatile acids + CO_2)	9.0 ± 0.8	9.3 ± 0.5	8.5 ± 0.5
Alkaline-ethanol extraction (oxalate)	6.5 ± 1.8	9.3 ± 1.0	12.0 ± 0.5

† Determined by titration to pH 6.1.

To conform with Eq. [1] and [2], D_P is defined by the diffusion equation $\partial pP/\partial t = (\partial/\partial x)[D_P(dpP/dx)]$ and may be shown to be given by $D_P = (2.303\theta f/b_P)(D_{LP}[Psoln])$, where b_P is the pP buffer power $-d[P]/dpP$. In most soils, over a considerable range of phosphate concentration, b_P is nearly constant i.e., the Temkin isotherm: $\Delta P = A\log[Psolution]+B$, where A and B are constants, is followed approximately.

The solution of Eq. [3] and [4] is somewhat complicated (Nye, 1983), and in general has to be achieved numerically. However, if phosphate occurs only in low concentration in the soil solution then $(\partial pH/\partial pP)_{HS}$ may approximate to zero, and an analytical solution is possible for constant concentration change of pP and pH at the boundary, which reveals the essential processes at work. For the calculation of changes around roots the radial coordinate forms of Eq. [3] and [4] are solved.

RESULTS

The effect of release of H^+ at a root surface on the concentration of P in the soil solution, as the interaction coefficient $(\partial pP/\partial pH)_{HS}$ increases, is shown in Fig. 6A. The change in the total amount of P around the root is shown in Fig. 6B. It will be noted that the release of H^+ can raise the concentration of P in solution above its initial value. Consequently, some P may diffuse away from the root, as shown in Fig. 6B.

THE RELATION BETWEEN THE AMOUNT OF H^+ RELEASED BY THE ROOT AND THE AMOUNT OF PHOSPHATE SOLUBILIZED

Figure 7 shows the form of the total P concentration profile when $D_{HS} = 2D_P$ and when $D_{HS} = 0.5D_P$. In each example, the amount of H^+ released into the soil and the value of the interaction coefficient λ is the same. The P concentration profile when there is no interaction

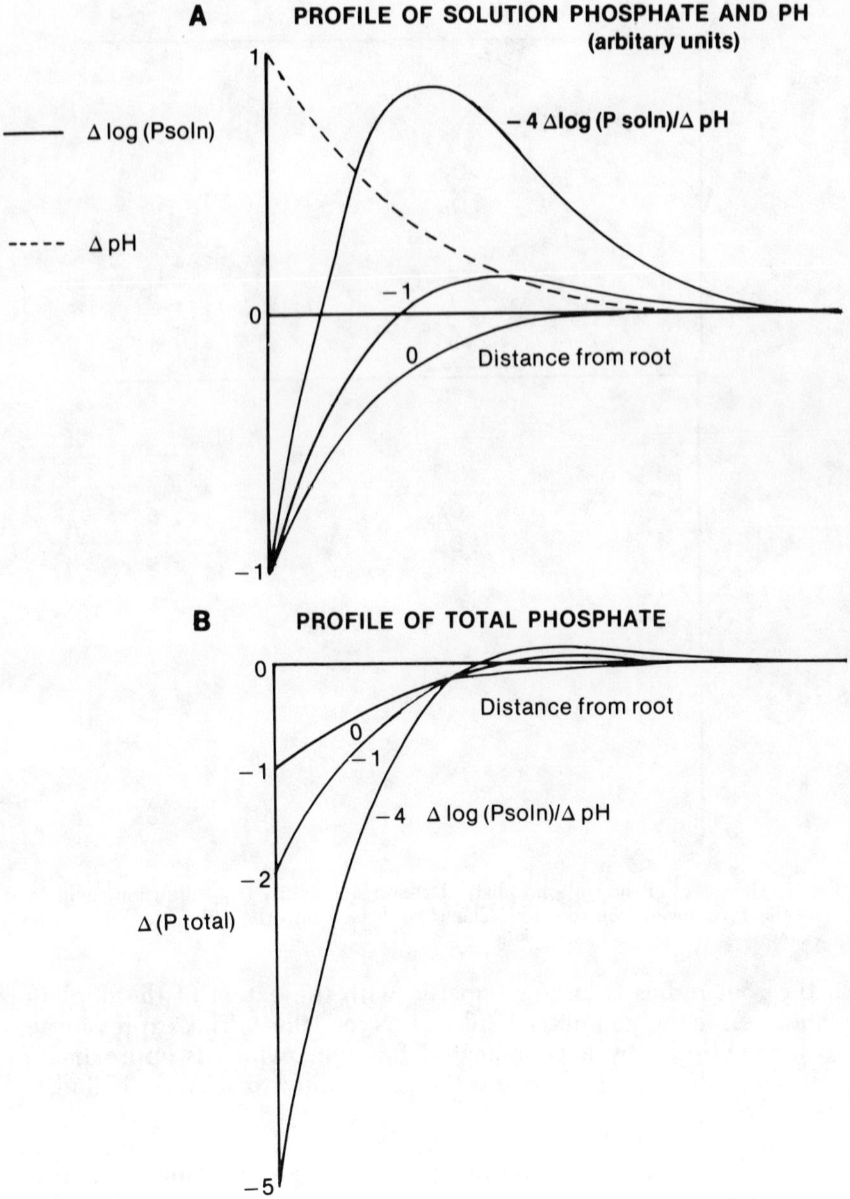

Fig. 6. A. Profile of soil solution phosphate and pH adjacent to plant root for different degrees of interaction. B. Corresponding change in the total P in the soil.

($\lambda = o$) is also shown. It may be shown by integration that the extra P absorbed by the root as a result of H^+ excretion is given by the formula

$$\Delta P \text{ total}/\Delta HS \text{ total} \simeq -(\partial pP/\partial pH)_P \, (b_P/b_{HS}) \,/[1+(D_{HS}/D_P)^{1/2}], \quad [5]$$

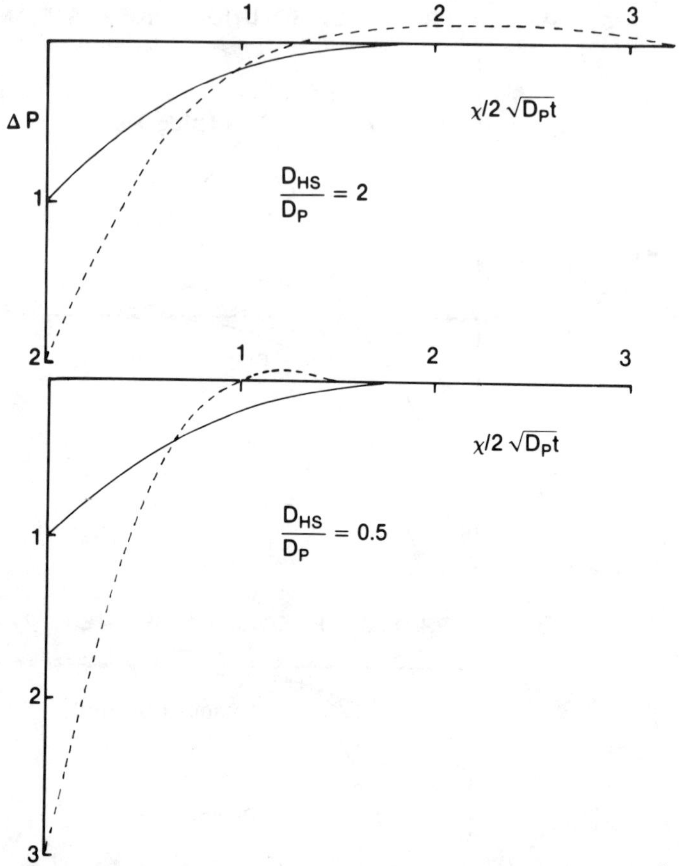

Fig. 7. The effect of the soil phosphate (D_P) and soil acidity (D_{HS}) diffusion coefficients on the P removed from the soil. The H^+ released into the soil is the same in both examples.

if the root radius is large compared with the extent of the depletion zone; i.e., if the geometry is planar (Nye, 1984). This expression has to be multipled by a complicated function, which is approximately $1/[(\pi D_{HS}t)^{1/2}/4a + 1]$ where a is the root radius, to allow for cylindrical geometry.

The first part of the formula [5], $-(\partial pP/\partial pH)_P(b_P/b_{HS})$, is the same as the phosphate that would be solubilized (i.e., the amount that has to be removed from the soil to maintain a constant pP) by a uniform addition of H^+ to a sample of soil. The second part shows the effect of diffusion. If D_{HS} is much less than D_P the P solubilized lies close to the root and nearly as much extra P is absorbed as would be solubilized in a batch of soil. If D_{HS} is much greater than D_P the P is solubilized at a much greater distance from the root and much less extra P will be absorbed. Of course, this argument assumes the root is in an infinite volume of soil. As the rooting density increases, the more rapidly will

P that has been solubilized and moved away from one root be recovered by adjoining roots. It is important, therefore, to assess the relative magnitudes of D_{HS} and D_P.

THE MAGNITUDE OF D_{HS} AND D_P

We have seen that D_{HS} and D_P are defined by

$$D_{HS} = (2.303\theta f/b_{HS})\,(D_{LH}[H_3O^+] + D_{LC}[HCO_3^-]) \quad \text{and}$$

$$D_P = (2.303\theta f/b_P)\,(D_{LP}[\text{P soln}])\ .$$

Typical values of b_{HS} range from 10^{-2} to 10^{-1} mol L^{-1} pH^{-1}, and of b_P from 10^{-4} to 10^{-3} mol L^{-1} pP^{-1}. Also $D_{LH} \simeq 8 \times 10^{-5}$ cm^2 s^{-1}, while $D_{LP} \simeq 0.8 \times 10^{-5}$ cm^2 s^{-1}. Hence at pH = 5 (where $[HCO_3^-]$ is negligible) and pP = 5, D_{HS} is likely to be an order of magnitude less than D_p. In these "average" conditions the root would absorb nearly all the P potentially solubilized by the H^+ it released. It is evident, however, that it would be possible to choose conditions, for example with high concentration of H_3O^+ and low concentration of P in solution, in which D_{HS} would be greater than D_p and the root would absorb little of the P potentially solubilized.

CONCLUSIONS

The equations describing the simultaneous release of H^+ from a root and its absorption of any phosphate solubilized have been formulated, and examples of their solution given. In most cases, numerical computer solutions are required for an accurate description of the development of the P concentration profile around the root. However, an adequate estimate of the extra phosphate a root may take up if it releases a known amount of acid should be given by the simple formula presented. We need to know much more about the factors controlling the rate of acid or base release from roots, and also the consequent effect of pH change on phosphate solubility over a wide range of soils.

REFERENCES

1. Bagshaw, R., L.V. Vaidyanathan, and P.H. Nye. 1972. The supply of nutrient ions by diffusion to plant roots in soil. VI. Effects of onion plant roots on pH and phosphate desorption characteristics in a sandy soil. Plant Soil 37:627–639.
2. Barber, S.A., and M. Silberbush. 1984. Plant root morphology and nutrient uptake. p. 65–87. *In* S.A. Barber and D.R. Bouldin (ed.) Roots, nutrient and water influx, and plant growth. Spec. Pub. 49. Soil Science Society of America, Crop Science Society of America, and American Society of Agronomy, Madison, WI.
3. Brewster, J.L., K.K.S. Bhat, and P.H. Nye. 1975. The possibility of predicting solute uptake and plant growth response from independently measured soil and plant char-

acteristics. III. The growth and uptake of onions in a soil fertilized to different initial levels of phosphate and a comparison of the results with model predictions. Plant Soil 42:197–226.

4. ----, ----, and ----. 1976. The possibility of predicting solute uptake and plant growth response from independently measured soil and plant characteristics. V. The growth and phosphorus uptake of rape in soil at a range of phosphorus concentrations and a comparison of results with the predictions of a simulation model. Plant Soil 44:295–328.
5. Cunningham, R.K. 1964. Cation-anion relationships in crop nutrition. III. Relationships between the ratios of sum of the cations: Sum of the anions and nitrogen concentrations in several plant species. J. Agric. Sci. 63:109–111.
6. Grinsted, M.J., M.J. Hedley, R.E. White, and P.H. Nye. 1982a. Plant-induced changes in the rhizosphere of rape (*Brassica napus* var. Emerald) seedlings. I. pH change and the increase in P concentration in the soil solution. New Phytol. 91:19–29.
7. ----, ----, ----, and ----. 1982b. Plant-induced changes in the rhizosphere of rape (*Brassica napus* var. Emerald) seedlings. II. Origin of the pH change. New Phytol. 91:31–44.
8. ----, ----, ----, and ----. 1982c. Plant-induced changes in the rhizosphere of rape (*Brassica napus* var. Emerald) seedlings. III. Changes in L value, soil phosphate fractions and phosphatase activity. New Phytol. 91:45–56.
9. Henry, P.S.H. 1939. Diffusion in adsorbing media. Proc. R. Soc. London 171A:215–241.
10. Hedley, M.J., P.H. Nye, and R.E. White. 1983. Plant-induced changes in the rhizosphere of rape (*Brassica napus* var. Emerald) seedlings. IV. The effect of rhizosphere P status on the pH, phosphatase activity and depletion of soil P fractions in the rhizosphere, and on the cation-anion balance in the plants. New Phytol. 95:69–82.
11. Israel, D.W., and W.A. Jackson. 1978. The influence of nitrogen nutrition on ion uptake and translocation in leguminous plants. p. 113–129. *In* C.S. Andrew and E.J. Kamprath (ed.) Mineral nutrition of legumes in tropical and sub-tropical soils. C.S.I.R.O., Melbourne, Australia.
12. Nyatsanga, T., and W.H. Pierre. 1973. Effect of nitrogen fixation by legumes on soil acidity. Agron. J. 65:936–940.
13. Nye, P.H. 1972. The measurement and mechanism of ion diffusion in soil. VIII. A theory for the propagation of changes of pH in soils. J. Soil Sci. 23:82–92.
14. ----. 1981. Changes of pH across the rhizosphere induced by roots. Plant Soil 61:7–26.
15. ----. 1983. The diffusion of two interacting solutes in soil. J. Soil Sci. 34:677–691.
16. ----. 1984. On estimating the uptake of nutrients solubilized near roots or other surfaces. J. Soil Sci. 35(3): (In press).
17. ----, and P.B. Tinker. 1977. Solute movement in the soil-root system. Blackwell Science Pub., Oxford, United Kingdom.
18. Riley, D., and S.A. Barber. 1969. Bicarbonate accumulation and pH changes at the soybean (*Glycine max* (L.) Merr.) root-soil interface. Soil Sci. Soc. Am. Proc. 33:905–908.
19. Smiley, R.W. 1974. Rhizosphere pH as influenced by plant, soils and nitrogen fertilizers. Soil Sci. Soc. Am. Proc. 38:795–799.

Chapter 6

Dependence of Crop Growth and Yield on Root Development and Activity[1]

DONALD A. BROWN AND H. DON SCOTT[2]

The relationships between root distribution soil and crop yields have been of prime concern to agronomic researchers for several years. A significant effort has been made to characterize the root distributions of agronomic crops in diverse soil conditions and at different sampling times during the growing season. In addition, efforts have been made to (a) determine soil conditions which contribute to optimum root growth and (b) modify soils for greater distribution of roots in the soil profile. The hypothesis behind this research has been that greater root exploration of the soil volume will lead to greater water and nutrient absorption and, thereby, reduce the risk of water and nutrient stress. Minimizing plant stress should, in turn, lead to increased plant growth and development and eventually to greater seed yields. The objective of this chapter is to present an overview of the factors affecting root distribution of agronomic crops and the significance of changes in root distribution patterns during growth as they relate to crop yield.

---

[1] Published with the approval of the Director of the Arkansas Agric. Exp. Stn. Appreciation is extended to the Arkansas Soybean Promotion Board for their support in portions of the research reported.

[2] Professors of soil chemistry and soil physics, respectively, in the Dep. of Agronomy at the Univ. of Arkansas, Fayetteville, AR 72701.

Copyright © 1984 Soil Science Society of America, Crop Science Society of America, and American Society of Agronomy, 677 South Segoe Road, Madison, WI 53711. *Roots, Nutrient and Water Influx, and Plant Growth.*

PROBLEMS IN RELATING ROOT DISTRIBUTION TO CROP GROWTH

There are two principal reasons why researchers have experienced difficulty in relating root distribution to crop yield: (a) the complications associated with the existence of a functional equilibrium between the growth of the shoot and root portion of the plant and (b) problems associated with the isolation and sampling of root systems under field conditions by nondestructive methods.

Functional Equilibrium

The functional relationship between the roots and shoots has been described by Brouwer (10). He described the functional equilibrium between the root and shoot system as an interrelated growth or cyclic growth pattern. In essence, the principle indicates that a significant change in the rate of shoot growth is transmitted to the roots and affects the rate of root growth; likewise, changes in the rate of root growth result in an effect on the shoot portion of the plant. Sanders and Brown (48) demonstrated this principle very convincingly by grafting supplemental roots or shoots onto soybean (*Glycine max* L. Merr) seedlings. Variations in the shoot:root ratio were obtained by grafting seedling soybeans with multiple root systems and/or multiple shoots. Plants with three roots on one shoot, three shoots on one root, and all combinations in between were compared with normal soybeans. In both controlled growth chamber and field experiments and regardless of the initial shoot:root ratio, the resulting shoot:root ratio of all plants at maturity under the conditions of these experiments was constant at 11:1 (Table 1). Similar results at maturity also were shown by West (60) for cotton (*Gossypium hirsutum* L.) grown in the field with varying ratios of supplemental shoots and root systems.

In both field studies the plants with multiple root systems outyielded the plants with single root systems by 28 and 53% for soybeans and cotton, respectively. All yield increases were due to an increase in the number of pods or bolls retained at maturity. These data indicate that an increased amount of actively growing roots provided sufficient amounts of water and nutrients to decrease plant stresses during the

Table 1. The effect of supplemental roots on the plant and yield characteristics of Lee 68 soybeans (48).

Shoot: Root ratio	Shoot wt.	Root wt.	Shoot: Root ratio	Leaf area plant	Seed wt. plant	Seeds/ plant
	—— g	——		cm^2	g	
1:1	110.75	10.99	10.12	7 846.45	81.73	539
1:2	130.87	11.76	11.05	9 437.47	104.69	718
1:3	172.97	15.56	11.20	12 152.59	108.20	724
$LSD_{0.05}$	36.49	3.32	NS†	3 098.08	20.83	143

† LSD = Least significant difference, NS = Not significant.

reproductive growth. This reduced stress in turn allowed the plants to set and retain more flowers and to mature more pods or bolls. Thus, the plants did not have to expend photosynthetic energy to build a larger root system and could devote more energy to building larger above-ground vegetative and reproductive systems. These results suggest that soil and crop management systems that minimize the amount of photosynthetic energy the crop must expend on a root system should lead to more energy available for top growth and to higher economic yield.

Measurement Techniques

The existence of a functional equilibrium between shoot and root growth gives rise to the need to observe the activity and development of the root system as clearly as the growth functions of the shoot have been observed. The basic problem has been to find a technique that will permit the study of crop root systems in situ. Development techniques for isolating plant root systems have been attempted by scientists for many years. Böhm (9) and Arkin and Taylor (5) have given extensive reviews of the methods which have been used during the last 10 years. The methods can be divided into destructive and nondestructive (in situ) procedures.

Destructive methods are those which necessitate removal of soil cores or the excavation of trenches or pits to sample plant roots from the experimental plot. These methods have been the most widely used but have made the continuous monitoring of root growth of a given plant throughout the season difficult if not impossible. They also destroy portions of the soil profile or the plant within the field plot.

Visual Observations. The first nondestructive techniques for relating root and shoot growth in the USA came with the use of the Rhizotron by Taylor in 1971 (56). Visual observations and photographing were recorded from underground soil bins covered with a transparent window, which allowed for observations of the root distribution and rate of root growth under field conditions. In 1974, Böhm (8) described the use of a periscope or mini-rhizotron for root observation underground. In 1978, Sanders and Brown (49) modified this method by the use of a medical duodenoscope equipped with a flash unit and a 35 mm camera. This unit was lowered into a square transparent tube set at a 45° angle in the ground and used to photograph roots growing across the tube walls.

These mini-rhizotron methods have added a new dimension to root research. They permit a continuous monitoring of root growth with soil depth, across the furrows, and throughout the growing season without soil or plant destruction. Simultaneously, the growth parameters of the shoot portion of the plant can be measured throughout the season. Recent improvement in this method includes the replacement of the 35 mm camera with a television camera and video recorder system for recording root photographs to give the mini-rhizotron method additional advantages in the time, labor, and number of observations that can be made during the season.

Sanders and Brown (49) found the scope method to be much faster than the core method. Estimates of the time required for root measurements on plants in a single location by taking core samples (eight replicate cores at two sites) to a depth of 0.72 m approximated 8 man h/sample. This included the proportionate time required for installation of the tube, taking the photograph, and calculation of the data for a sample. Böhm (9) compared the time required for characterizing plant roots by five methods. He reported a minimum time of 40 min for the mini-rhizotron and 26 h/sample for the framed monolith and pinboard method. The core-sampling method required 9.5 man h/sample. The cost of equipment for the scope and 35 mm camera ranges from $3000 to $6000. The cost of a custom-made scope which uses a micro-size (0.038 × 0.127 m) portable color television camera with accompanying batteries, light source, tape recorder, and monitor is about $15000 (private correspondence with Dan Upchurch, USDA, ARS, Lubbock, TX).

Even with the mini-rhizotron method, a significant amount of research is still needed to refine the procedure in terms of the number and location of observation tubes, number or frequency of measurements, and the statistical treatment of the data. There also remains a question of the resolving power of the scope in terms of its ability to measure differences between root distribution of contrasting cultivars. Refinements in the procedure will continue and there is every indication that the use of mini-rhizotrons will enable scientists to make significant progress in understanding the relationships between root growth and crop yield.

Porous Membrane-Root Culture Technique. The use of a porous membrane envelope technique represents the most recent attempt to measure root and shoot growth under field conditions (11, 12). This technique enables the roots to grow inside the membrane envelope and to absorb nutrients and water from the soil solution without interference from the variable physical and chemical properties within the soil profile. It also permits the harvesting of the whole plant with its root system intact and clean (Fig. 1). This method was successfully tested in 1982 using soybeans.

Briefly, this method involved growing soybean plants in the field with their root system restricted to the inside of a porous, permeable membrane envelope (Gelmans Inc., vescore membrane with a pore diameter of 3×10^{-6} m). Membrane envelopes measuring 0.30 m wide and 1.50 m long were prepared, sealed on the sides and lower end, and left open at the top.

The membrane envelope was flushed with a suspension of N_2-fixing inoculent, and laid horizontally at the 0.20-m soil depth in a conventionally prepared bed and covered with soil; the open end was left protruding at the soil surface (Fig. 1). A 2-week-old soybean seedling was transplanted in its original container (with the bottom removed) into the open end of the membrane envelope. The plant was staked for

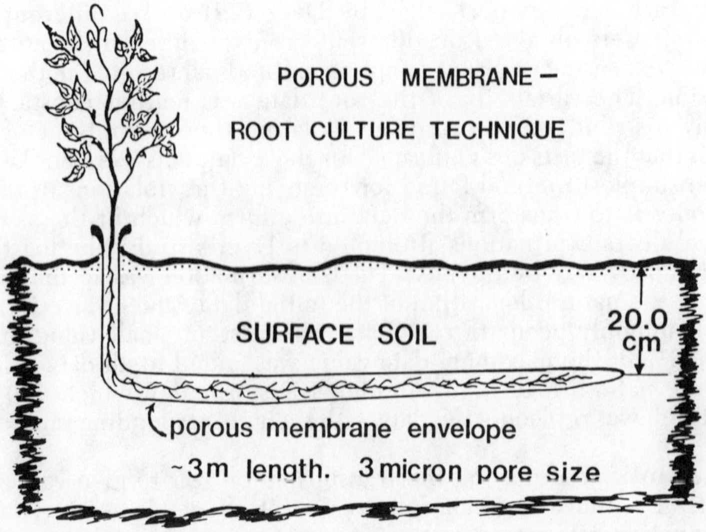

Fig. 1. A porous membrane-root culture technique designed to grow soybeans in the field under minimal soil stress conditions. Brown and Haq, 1983 (11).

support, and a water basin was constructed over the membrane area. The plant was supplied with nutrient solution for 1 week; afterwards, all nutrients were derived from the soil solution. The soil was fertilized uniformly with 224 kg/ha each of P and K at planting with an additional fertilization of 224 kg/ha mixed in the surface 0.1 m at the R-1 stage of growth. Water, applied twice a week kept the soil and membrane above the 30-kPa potential throughout the season. Five replications and six sampling dates (V8, V12, R1, R3, R5, and R8) were repeated for three contrasting cultivars (R80–84, R80–61, and R74–522). At each sampling stage, the plant and its roots in the membrane were removed for physical and chemical analyses. Shoot parameters were measured at each stage of growth and related to root growth and yield.

Statistical Analyses. Along with the difficulties in growing and retrieving roots for study, problems associated with plant root sampling also include statistical analyses of root data, along with spatial variability within the soil-root environment and plant canopy.

The analysis of variance tests have been routinely used to analyze root distribution data in order to interpret and identify significant treatment effects. The analysis of variance (ANOVA) test is based on three assumptions that must be met by the data if the test results are to be valid. Two of these assumptions are relative to the population of values at a single level of all treatments (i.e., replications). First, the average of the differences between individual replications and their mean is assumed to be zero. Second, the replications are assumed to be normally distributed about their mean. The third assumption regards the data population at all levels of the treatments as a whole. The variance among replications is assumed to be constant throughout the data.

Separate analyses performed by Dyer (22) on six different root distribution sets obtained on different crops at different locations in different years using different sampling methods all resulted in the same conclusion. The variability of the root data was neither constant nor normally distributed. The result of violating the assumptions of normality is that the tests of significance for those data sets were not known.

The simplest method found for removing the violations of the assumption was to transform the data into a form which fit the assumptions. Of all transformations attempted in Dyer's study, the log transformation proved to be the best. The transformation was accomplished by taking the natural logarithm of the initial data value. Since 0 (zero) values commonly occur in root data, a consistent small value on the order of 1% of the maximum data value was added to all data prior to taking the natural logarithm. Another transformation which also performed well was replacing the data with their corresponding rank-order value.

The purpose of making the transformation was to give validity to the tests of significance regarding treatment effects. When the data in the original six data sets were transformed into an acceptable form, the interpretation of the statistical analyses was usually changed to some degree. The determination or nonsignificance of specific treatments was changed in several instances after data transformation. Because the violation of the assumptions by root distribution data in the raw form can often lead to misinterpretation of the results of the experiment, it was strongly recommended that all statistical tests be performed on log or rank transformed roots distribution data. Soybean residual distribution curves are shown in Fig. 2 for untransformed (root length density, RLD) and transformed data (Log, Rank).

Root distribution also has a significant variation associated with plant age. Root distribution measured at only one time during the growing season can lead to an incomplete or erroneous interpretation of the relationship between root distribution and crop yield. Sequential measurements must be taken to give meaning to changes in total root growth as well as changes in root distribution.

Variability Sources. Problems in the statistical analyses of root data are further complexed by variability in the soil and plant environment. Spatial statistics have recently been used to eliminate this problem. Spatial variability within the plant root-soil environment has presented formidable problems since soil is a complex and heterogeneous porous media with widely varying properties. As a result, it is not uncommon within a field to find several soil mapping units, each having a range of soil properties such as designated horizons, profile thicknesses, textures, structures, and pH and nutrient levels. Perhaps even greater field variability occurs with the more dynamic soil-water properties of hydraulic conductivity, water retained, and water potential. Since the magnitude of each of these properties at one location is usually related to its magnitude at another location in the same field, the use of classical statistics which assumes spatial independence is not appropriate. As a result, a special branch of statistics, called geostatistics, was developed

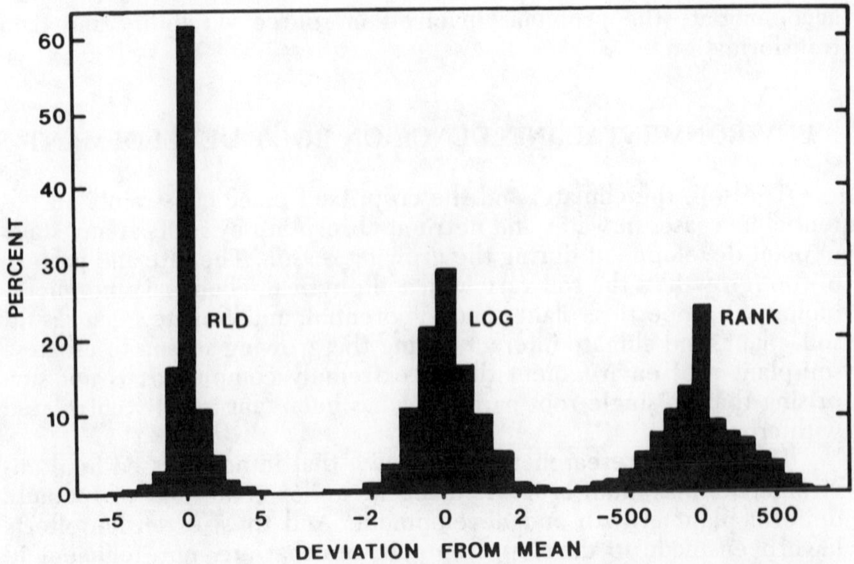

Fig. 2. Comparison of soybean root residual distribution; root length density (RLD), untransformed; Log, natural log transformation; and Rank transformation. Dyer, 1982 (22).

by Matheron (35) to solve problems that deal with spatially dependent parameters. This relatively new statistical tool being applied to soil variability uses structural analyses such as the autocorrelgram and/or varigram to determine the spatial relationships of a regionalized parameter. The distribution of observations over the field can be described by an extrapolation process called "kriging" (11, 14) which also can be used to estimate the error of interpolation. Examples of recent research reports which use these techniques to determine the spatial variability of soil properties include Vieira et al. (57), Gajem et al. (27), Burgess and Webster (14, 15), Webster and Burgess (59), and Hajrasuliha et al. (29).

Fewer studies have been conducted on the field variability of the crop canopy. Scott et al. (53) have shown that the dry weight of a determinate soybean grown in a relatively uniform field varied with plant part, node, and sampling time. Coefficients of variation (CV) of the whole plant were relatively high at the lower nodes, decreased to node 12, and then increased toward the top of the plant. Values of CV at the bottom nodes were lowest for stem segments, and variability of the leaves and petioles were greater due to abscission. The relatively high variability in the upper nodes of all plant parts was attributed to differences in initiation of new nodes with attendant leaves and petioles. These results indicate that the heterogeneous nature of field soils cause a high degree of variability in root growth which transmits a high degree of variability in the growth characteristics of the top portion of the plant. Any analyses of root systems or shoot:root relationships must accommodate the problems involved in source variability and data transformation.

ENVIRONMENTAL INFLUENCE ON ROOT DEVELOPMENT

The soil, the climate, and the crop itself place constraints on potential increases in water and nutrient absorption by roots at any stage of plant development during the growing season. The rate and pattern of root growth in the soil vary with soil physical, chemical, and microbiological properties; plant genetic potential; and climate. As a result, soil, plant, and climate interact during the growing season to create a soil-plant root environment that is extremely complex. It is not surprising that no single root parameter has been consistently correlated with crop yield.

Historically, researchers have shown that in nonstressed field environments maximum crop yields occur under conditions of spatially uniform plant growth and development. And thus, extensive efforts have been made to develop crop cultivars that are more efficient in their utilization of available resources such as water, nutrients, CO_2, light, oxygen, and perhaps space. Since crops extract practically all their water and nutrients from the soil, the volume of soil explored by the roots is imporant. This importance is illustrated in Fig. 3 for a 'Lee 74' soybean grown on a Crowley silt loam (fine, montmorillonitic, thermic Typic Albaqualf) in rows spaced 0.8 m apart. For an average soybean plant the roots will be distributed in the x, y, and z planes in a manner that depends upon the interactions between the crop genetic potential and the soil and climatic environments. Research in Arkansas has shown that for a well-watered (without water stress) Lee 74 soybean plant the approximate seasonal (120 days) requirements are 16.5 L of water

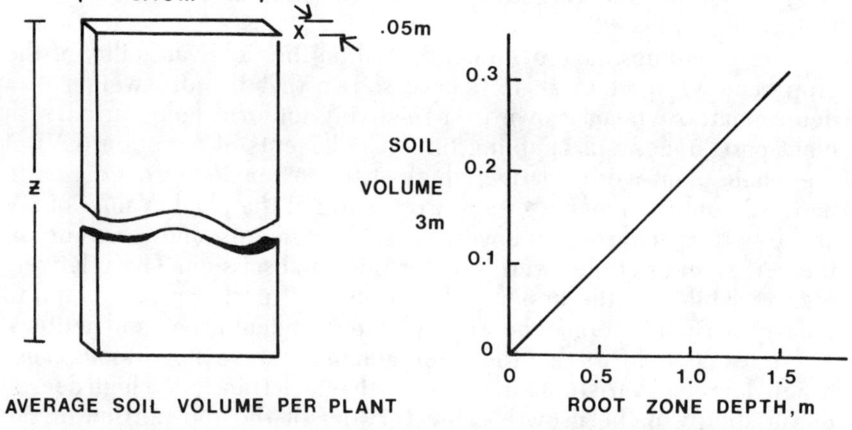

Fig. 3. Relationship between depth of root zone and volume of soil explored by the roots.

(transpiration only) and 0.80, 0.69, 0.53, and 0.14 g of N, K, Ca, and Mg, respectively.

The need for water and nutrients is continuous throughout the growing season. For the well-watered Lee 74 soybeans this need is illustrated in Fig. 4 by the similar seasonal patterns of cumulative extraction of water and K and of dry matter produced. These cumulation curves can be subdivided into three phases. The first can be designated as the exponential phase and usually lasts from emergence to approximately the first 25 days. In this phase, the plant grows vegetatively and has high relative growth rates of both roots and shoots. Extension of the roots in the soil is greater in the z direction than in the y direction. The second phase can be designated as the linear phase and usually lasts for approximately 80 days. Both vegetative and reproductive growth occur during this phase. Root and crop growth rates are highest, and the greatest demands for water and nutrients are placed on the roots during this period. Roots continue to extend downward to "fill in" the soil volume near the soil surface. Simultaneously, the roots extend in the y direction toward the row middles. The third phase can be designated as the constant stage and occurs from the time when the crop has reached maximum dry matter accumulation until harvest. Both roots and shoots begin to senesce and die during this period, and little demand is placed on the roots for water and nutrients. Similar relationships can be shown for other nutrients and other crops grown under well-watered conditions in other years.

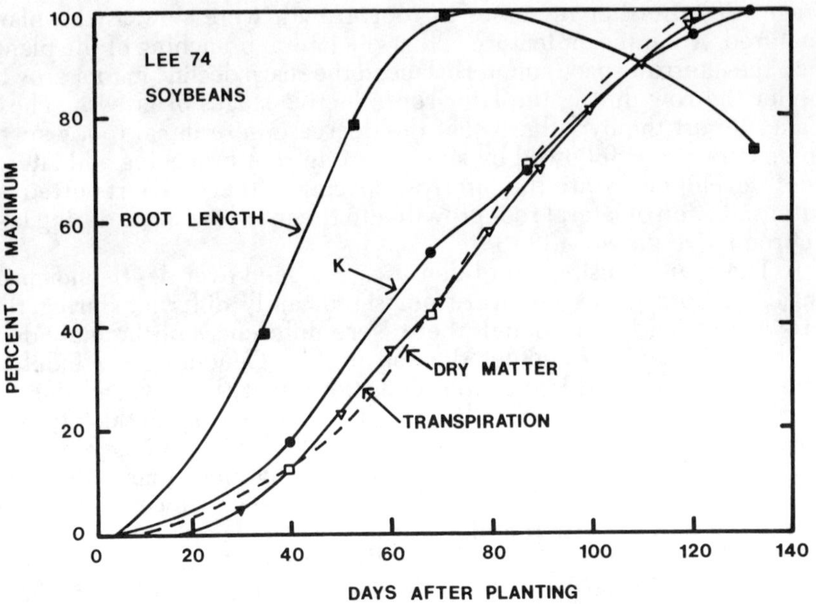

Fig. 4. Relationship between percent of maximum and days after planting for cumulative transpiration, dry matter, K assimilated, and root length of well-watered Lee 74 soybeans grown on Crowley silt loam.

Crop Root Distribution

Extensive research data on root growth has been published. These reports have primarily been concerned with the effect of specific soil parameters on root growth, including soil moisture (4, 36, 47, 54), soil strength (2, 20, 51), soil aeration (32, 24, 43, 44), temperature (3, 18, 40, 41, 47), crop rooting (6, 7, 33, 35, 55), and modeling program (19, 39, 42, 58).

The root distribution for soybean, cotton, and corn (*Zea mays* L.) were presented here to illustrate the changes with plant age, genetic characteristics of the crop, soil depth, soil moisture, and plant competition. Emphasis is given to the effect of these complex influences upon crop growth and yield.

Soybeans. The measurement of root distribution of soybean by means of the fiber optic scope was reported by Sanders and Brown (49) in 1978. Changes in root density for the determinate soybean cv. Lee 74 and an indeterminate cv. R72–346 were reported. From this study, the changes in the distribution of roots for five physiological stages of soybean growth by Fehr and Caviness (25) are shown in Fig. 5. Root densities in the row for both determinates and indeterminate genotypes increased linearly through the mid-vegetative period (V8) and then increased exponentially to a maximum in the early reproductive stage (R3). This was followed by a sharp decline in root density to maturity (R7). Significant growth of roots into the interrow space (0.25 and 0.50 m from the row) did not begin until the mid-vegetative stage (V8) after which root density increased exponentially until the early reproductive stage (R3). Further increases in root growth were slow as the plant matured. A significant feature is that the lateral branching of the plants into the interrow space counterbalanced the sharp decline in root growth under the row during the later reproductive stages of growth. These data support the hypothesis that the decrease in terminal root growth under the row is followed by an increase in root branching and lateral root development into the interrow spacing. Others have reported an intensification of lateral root growth into the interrow space during late reproductive growth (33, 38).

The root intensities (root densities averaged over depth and spacings) for both genotypes were not significantly different during the growing season, even though there were differences in the individual root distribution throughout the soil profile. Generally, the indeterminate soybeans had higher root densities under the row, and the determinate soybeans tended to have higher root densities in the interrow space.

Determinate soybeans yielded significantly more than indeterminate soybeans (3078 and 2171 kg/ha, respectively). However, the only significant difference in root distribution was the higher interrow densities in determinate varieties during the R4 and R7 stages of soybean development. These data suggest that root distribution and activity may be more important than root intensity in influencing soybean yields.

Soybean root growth has been shown to be most under the row during the vegetative stage but gives way to a rapid lateral root growth

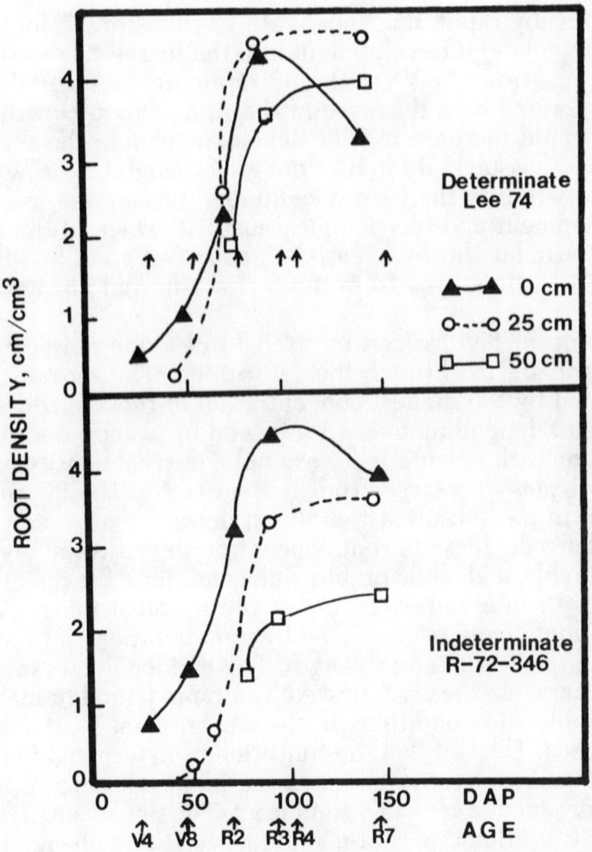

Fig. 5. Changes in the root density of determinate (Lee 74) and indeterminate (R72-346) soybeans at three distances from the row, 0, 0.25, and 0.50 m. Sanders and Brown, 1979 (48).

in the furrow during the reproductive stages. For example, the percentage of the total root system developing into the furrow was 24% during the V4 stage, 29% in the V8 stage, 47% in the R2 stage, 57% in the R4 stage, and 67% in the R7 stage of maturity. The roots under the row showed distinct signs of decay during reproductive growth while root extension into the furrow continued. This lateral root extension places an added emphasis on the need for adequate soil moisture in the furrow to maintain active root growth. Over 50% of the roots were distributed in the surface 30 cm of soil. Nonirrigated soybeans produced fewer roots but had a greater percentage of their roots deeper in the profile. In response to irrigation, the soybeans produced more roots evenly distributed between the rows and under the row.

Mitchell and Russell (38) described three stages of root growth. In our study, four distinct stages of root growth were related to the physiological age of the plant; (a) the V8 vegetative stage of top growth was characterized by rapid and linear rate of root growth under the row with minimum lateral development into the furrow space; (b) the transitional change from the V8 to the R2 reproductive stage of top growth was characterized by a decrease in the rate of root growth under the row and a rapid increase in root development into the furrow; (c) in the reproductive stage, R2 to R4, root growth under the row decreased, but root growth into the furrow continued to increase; and (d) in the seed development and physiological maturity stage of the plant (R7), root growth under the row ceased and showed a net loss due to root decomposition. Root growth in the furrow showed no significant new growth.

A significant feature can be used to describe most soybean root distribution patterns through the soil profile. The root distribution is characterized by the highest concentration of roots in the Ap horizon (15 to 40 cm, depending on soil) followed by a sharp decrease in root concentration with soil depth; for example, the root density for irrigated Lee 74 soybeans decreased from 6.5 cm/cm^3 at the 20 cm depth to 0.5 cm/cm^3 in the subsoil at the 50-cm depth.

This sharp decrease in root concentration marks the boundary between favorable and unfavorable soil conditions for maintaining the normal growth rate and extension of the terminal roots. Unfavorable conditions could have been caused by poor aeration and nutrient contents or by high acidity and strength. This sudden decrease in terminal growth is characteristically followed by a rapid lateral branching in the more favorable soil conditions in the Ap horizon, particularly in the interrow space. The fact that the transition from terminal to lateral root growth occurs in many instances as the plant is changing from vegetative to reproductive growth is significant. The slowdown in root growth during this transitional period may be a result of a change in the partitioning of the photosynthetic energy in favor of the initiation of reproductive parts. Thus, the functional equilibrium between the roots and shoots has been shifted so that bloom set and retention are affected in varying degrees; the response of a plant to a stress during this period would depend upon the intensity and duration of the stress as well as the ability of the plant to tolerate the stress.

Four soybean cultivars were evaluated for their genetic tolerance to moisture stress in the R2 and R4 stages of growth by Brown et al. (11). The cultivars studied included 'Sohoma', which has been reported to have drought tolerance in field tests in Oklahoma, and three other cvs. Lee 74, Davis, and Centennial, which have not been noted for tolerance to drought conditions. Root density measurements were made with the fiber optic scope at three growth stages (V13, R5, and R7).

The pattern of root development for Lee 74 and Sohoma grown in 1981 under moisture stress at the R4 stage is shown in Fig. 6. The effect of moisture stress at the R4 stage on Lee 74 root growth resulted in a root system decrease in root density from 7.4 to 5.2 cm/cm^3 in the 0 to 16 cm soil depth, with no differences shown in root densities at greater soil depths. The Sohoma cv. showed less effect from the moisture stress, with root density decreasing from 6.3 to 5.1 cm/cm^3.

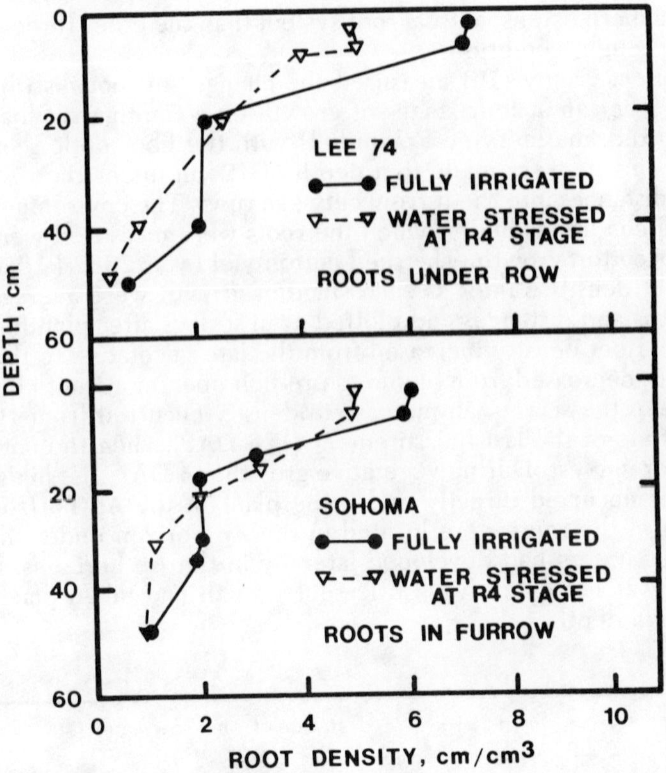

Fig. 6. Root distribution patterns for Lee 74 and Sohoma soybean cultivars at the R4 stage of growth. Brown et al. 1980 (13).

The overall root density data were not significantly different between the Lee 74 and Sohoma cultivars. Sohoma outyielded Lee 74 by 2361 to 1929 kg/ha—a difference not attributed to root distribution. Root density measurements under the row and in the middles showed that about 78% of the roots were in the 0 to 13 cm depth at stage R5 and about 75% at stage R7. Also, about four times as many roots were under the row as were present in the middle at both stages.

The overall root density of the irrigated soybeans was significantly higher than that for plants grown under moisture stress. Soybeans under full irrigation increased root density under the row and in the middles in the 0 to 16-cm depth at both the R5 and R7 stages. Davis produced over three times as many roots as the other cultivars in the 48 to 64 cm depth of the profile under the row when stressed during the R4 stage. Davis also produced a significantly higher root density than Sohoma and Centennial in the 16- to 32-cm section of the profile in the middles when moisture stress was applied at the R2 stage of growth. In 1980, which had limited rainfall, Sohoma produced more than twice as many roots in the middles as the other three cultivars. There was, however, no indication from the root growth data that Sohoma produced a more extensive total root system than the other three cultivars under drought conditions.

Cotton. Cappy (16) measured the changes in root distribution of Rex 713 cotton at four stages of growth on a Captina silt loam (fine-silty, mixed, mesic Typic Fragiudult) with the fiber optic scope (16). Measurements were made to a depth of 72 cm under the row and at 25 cm distances into the furrow between rows. The environmental and cultural conditions under which the roots were measured were favorable for cotton growth. The seed cotton yield was 2882 kg/ha.

Root density values at each stage of growth were averaged over all depths and distances and plotted against days after planting (DAP) (Fig. 7). Root density increased from the late vegetative to the bloom stage but decreased from bloom to pre-boll opening. Due to high rainfall late in the season, an increase in density occurred from 102 DAP to the time of the last measurement (126 DAP) when the cotton was ready for harvest. During vegetative growth, 44 DAP, the highest root densities occurred directly under the plant in the Ap horizon. Fifty percent of the roots were located in the Ap horizon under the rows. The root system had developed laterally in the Ap horizons, but this growth was not extensive. Little root growth had taken place below the 22-cm depth.

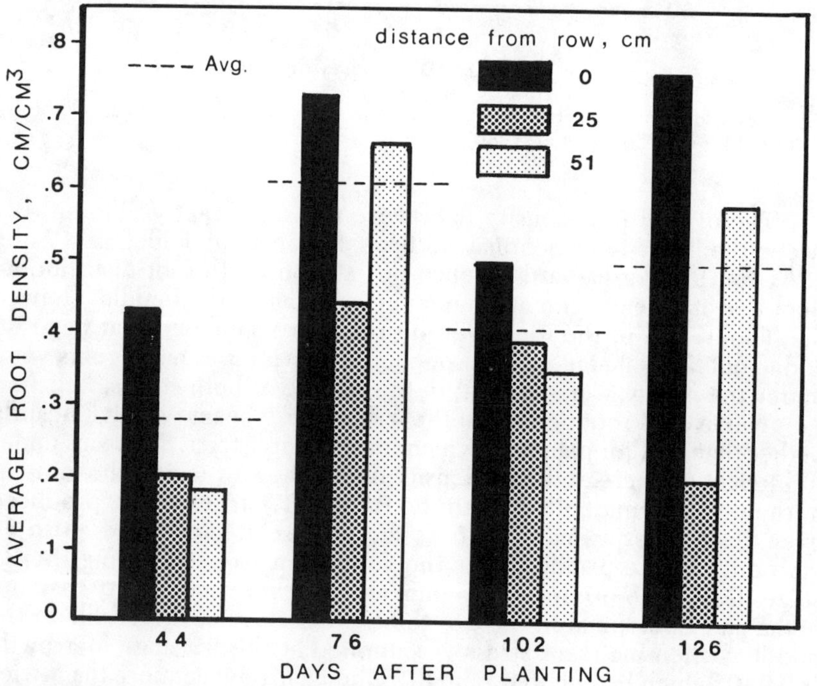

Fig. 7. Changes in the root distribution of Rex 713 cotton at four stages of growth. Cappy, 1979 (16).

Root measurements made 76 DAP, during the bloom stage of cotton, showed that most of proliferation in the root system took place during the 44- to 76-day interval, the root system essentially doubling in size. The data show that 83% of the increase in root density could be attributed to proliferation in the Ap2, B1, and B2t horizons. The highest root densities were found at 0 and 0.51 m from the center of the row. The increase in root density at 0.51 m over the 0.25-m distance at this stage of growth was attributed to the influence of two adjacent rows of cotton. The root systems of plants from each row can be expected to grow laterally toward the middle of the furrow. Thus, the location most likely to exhibit the influence of the adjacent rows in this study would be 21 cm from each of the rows.

During the period from 76 days to mid-bloom (102 DAP), the cotton plant sets fruit and the bolls begin to mature. At this particular stage the average root density (overall depths and distances) was 0.41 cm/cm^3, a decrease of 32% from the bloom stage. The decrease in most other horizons suggests that the rate of root loss was more rapid than that of new root growth.

Twenty-four days elapsed from the pre-boll opening measurement until the final root measurements were made 126 DAP. During the period, rainfall and other conditions were optimal for regrowth. Shoot regrowth was reflected in new root growth. The majority of the new root growth occurred in the Ap horizons under the plant (60%) and in the B1 and B2t (31%) horizons in the furrow (51 cm from the row). Root densities continued to decrease slightly in most other horizons.

Previous work on cotton root systems has suggested that considerable modification of the root system can occur under various soil conditions (1, 30, 45). Conditions such as wetting and drying cycles, subsoil acidity, and root restrictive layers in the soil affect root growth. Direct determinations of the cotton rooting pattern under field conditions have been lacking. Pearson (45) has reported a cessation of root growth at the time bolls mature but made no mention of possible regrowth. Bassett et al. (7), using a radiotracer technique, concluded that under favorable conditions in the field the root system never matures but continually develops as long as the cotton plant continues to grow. While the rooting pattern of Rex 713 cotton determined in this study is similar to that found in other research, the capabilities of the fiber optic technique has permitted a more comprehensive study, including distribution over depth, distance, and time, to give a more complete picture of the rooting pattern throughout the season. Because the measurements were of a nondestructive nature, the rooting pattern presented here follows the development of the same root system or systems throughout the entire growing season. Previously, nondestructive methods for measurement of undisturbed plant roots could not be made under field conditions.

Corn. The effect of seasonal climate on the root distribution patterns for corn was measured for 2 contrasting years by Carithers (17). The 1979 season was characterized by frequent light rains, but in 1980 the rainfall was inadequate for good corn yields and required supplemental irrigation.

These two contrasting seasons resulted in marked differences in corn root distribution (Fig. 8). The growth of corn roots increased rapidly in 1979 to a maximum at 75 DAP and then declined sharply through maturity. This rapid decline was due to moisture stress resulting from an inability to irrigate on the planned schedule. This effect on root growth was evident at all depths and within the furrow.

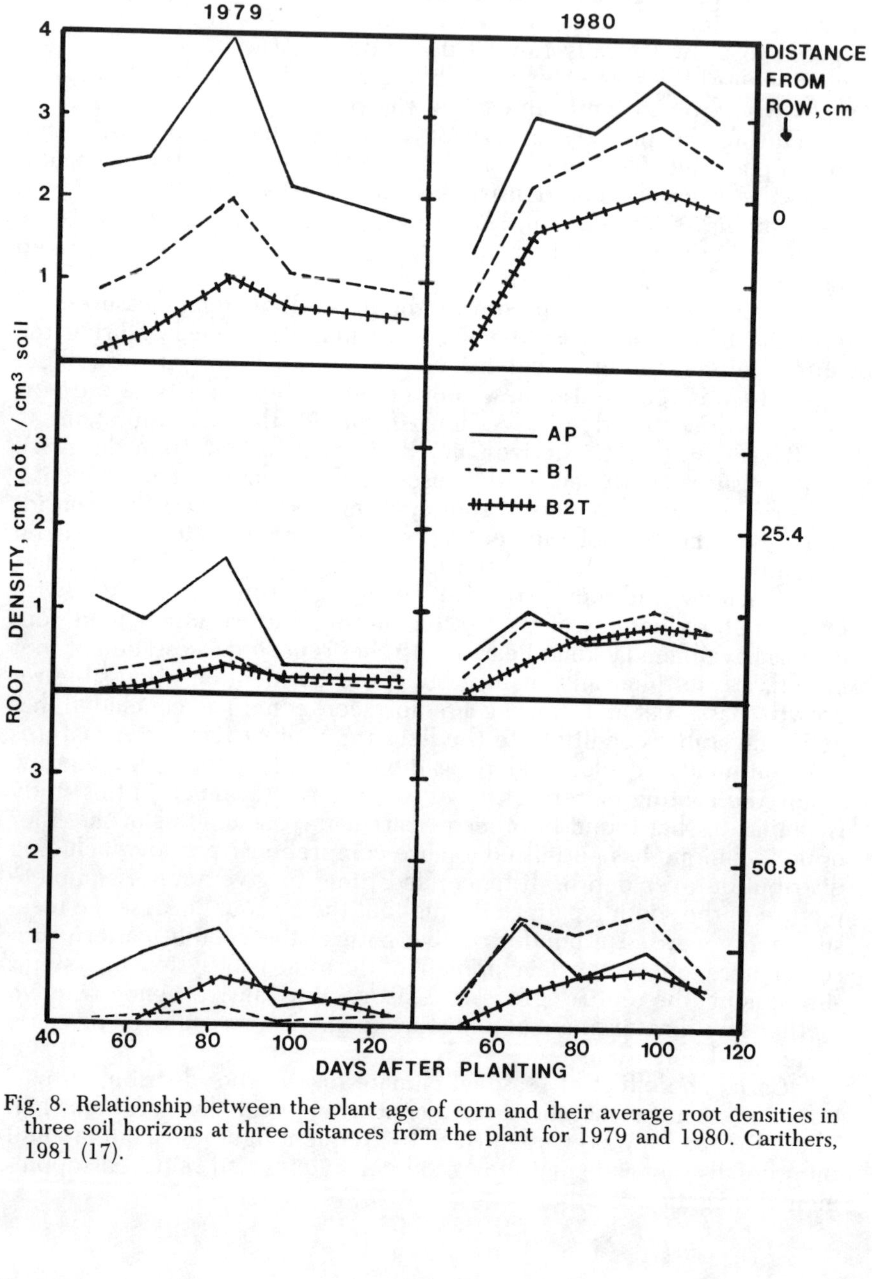

Fig. 8. Relationship between the plant age of corn and their average root densities in three soil horizons at three distances from the plant for 1979 and 1980. Carithers, 1981 (17).

In 1980, the rainfall was supplemented by scheduled irrigations. This resulted in extending root growth up to 100 days. The effect was shown on root growth at all soil depths and in the furrow.

In both seasons, root densities increased steadily reaching their maximum near the time of shift from vegetative to reproductive growth. By 70 DAP, 50 to 80% of the maximum total root density had developed. Most of this development occurred in the Ap and B1 horizons under the row with no significant root growth occurring in the B2t horizon until after this time. During the period when root densities were increasing, moisture stress often resulted in dehydration and senescence of roots especially in the Ap horizons in the furrow. Throughout the growing season, the relative distribution of roots according to lateral distance from the plant remained nearly constant; 60% under the row, 20% at the 25.4-cm distance, and 20% at the 50.8-cm distance. Soon after the corn plant went into reproductive growth, old roots began to senesce faster than new roots were formed, resulting in a net decrease in densities especially under the row.

The greatest water uptake was associated with higher root densities only when moisture was readily available. When soil water tensions were adversely high, regions of low root densities and higher water contents supplied most of the plant's water needs. The precise relationship between root density, soil water content, and water uptake in corn, as well as other crops, is one which needs further attention.

Effects of Plant Competition on Root Growth

The growth of crop roots within the soil profile can be severely limited by root competition for water and nutrients, particularly during the reproductive stage of growth. The deleterious effect of root competition on crop yield arises from two sources: (a) intraspecific competition resulting from high plant populations of a given species and (b) interspecific competition which arises from high populations of two plant species that compete with each other for water and nutrients.

Intraspecific Competition. The effect of intraspecific plant competition on soybean root distribution was measured by Cappy (16) for 'Forrest' soybeans grown under three-row spacings (0.51, 0.76, and 1.02 m) and two plant population densities (13 and 26 plants/m). Root distribution was measured at four stages of growth (V9, R2, R6, and R8).

The effect of row spacing and plant population on the distribution of roots at the R4 stage of growth for depth and distance from the row are shown in Fig. 9. Increasing the number of soybean plants from 13 to 26 plants/m in the narrow rows (0.51 m) resulted in a significant increase in root density under the row and in the furrow. The relatively high root density in the furrow at all depths was due to the fact that the supplemental moisture was applied by furrow irrigation. There was good movement of water into the soil under the furrow area, but the water did not adequately spread across the row and under the plants.

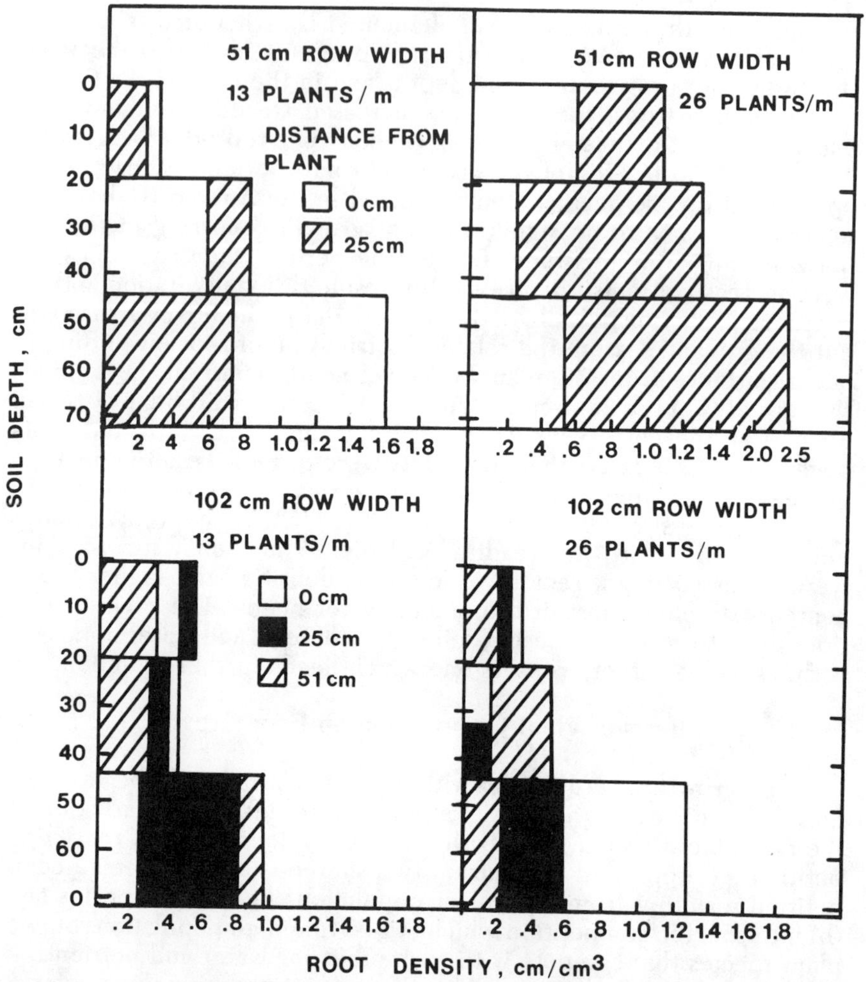

Fig. 9. Root distribution of Forrest soybeans at the R4 stage of growth; planted in narrow (0.51 m) and wide (1.02 m) rows at two population densities; 13 and 26 plants/m, and for two or three distances from the plant. Cappy, 1979 (16).

The effect of moisture distribution in the soil profile on root growth was evident.

Increasing plant population from 13 to 26 plants/m in the wide rows (1.02 m) did not alter root growth to the extent that it did in the narrow rows. The relatively higher root density in the subsoil was also due to a better penetration and distribution of soil moisture under the furrow.

The highest yields were obtained at the plant population of 13 or 16 plants/m on rows 51 cm apart (2028 and 2278 kg/ha) or on rows 1.02 m apart with populations of 26 plants/m (2164 kg/ha).

Two general conclusions concerning the effects of row spacing and plant density on root distribution can be made: (1) the closer the rows, the greater the root density in the furrow between the rows and (2) the greater the plant population, the greater the root density in the horizons directly under the row. This observation is in agreement with the work of Raper and Barber (46) who indicated that competition between plants limits the soil volume explored by the roots and induces primary lateral root growth. Sanders and Brown (50) presented data which indicated that taller soybeans with less branching have greater root densities beneath the row than in the furrow. Their data also suggest that more branched soybeans have greater root densities at the lower depths in the furrow.

Interspecific Competition. The effect of intra- and interspecific plant competition between soybeans (Lee 68) and tall morning glory [*Ipomoea purpurea* (L.) Roth] upon root growth, water and nutrient absorption, and soybean yields has been reported by Scott and Oliver (52).

This research provides data to support the concept that competition between different plants is related to the temporal and spatial differences in development of root systems and resulting differences in water and nutrient uptake. Figures 10 to 14 represent the results of a study on the interference of an aggressive weed, tall morning glory, and soybean grown under intra- and interspecific conditions on a silt loam in the field. Root samples were taken by the core method at three locations from the rows at 0.125-m depth intervals to a depth of 1.0 m and at four times during the growing season. The root densities obtained from a given location were summed over depth to give root intensities.

The relationships between root intensity and distance from the center of the row for soybeans and tall morning glory when grown under intraspecific conditions are shown in Fig. 10 and 11, respectively. These results show that the root intensities of these species vary depending upon the location in the row and plant age. Root intensities of soybeans were highest in the center of the row and underneath the shoot and decreased exponentially away from the row center. As the season progressed, soybean root intensities generally increased; however, little overall increase in root intensity was noted after 8 weeks after planting. This critical period corresponded to the onset of reproductive growth. Similar results also were noted with tall morning glory except that a greater proportion of its roots were found away from the row center and that a large increase in root intensity of tall morning glory occurred between 8 and 12 weeks after planting. This large increase was due primarily to an increase in root density at the greater depths of the profile. From these results it appears that during vegetative growth the determinate Lee 68 soybeans produced roots at rates similar to those of tall morning glory and would compete favorably for resources such as water, nutrients, and oxygen. During soybean reproductive growth, however, root intensity of the weed increased at a much greater rate than that of soybeans, and therefore, would be ex-

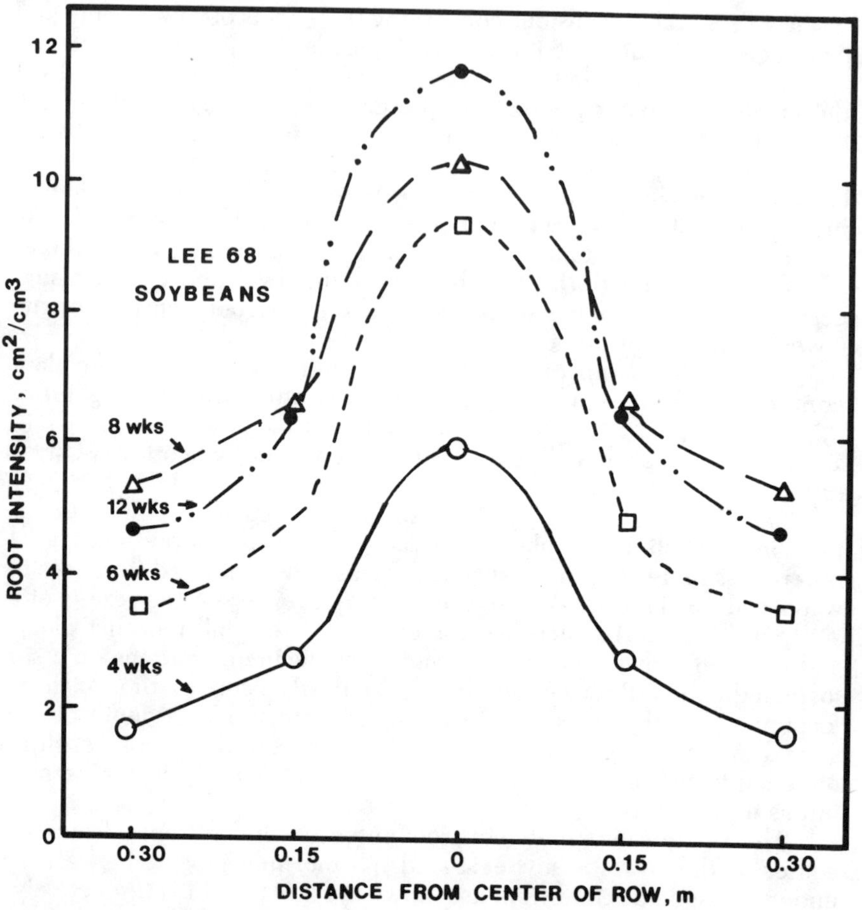

Fig. 10. Relationships between soybean root intensity and distance from row center at selected plant ages.

pected to have an advantage for the extraction of water and nutrients during the critical reproductive growth of soybeans.

The effects of intra- and interspecific competition (and/or interference) on the extraction of water from the soil profile are illustrated in Fig. 12, 13, and 14. Under intraspecific conditions, soybeans extracted a large portion of their water from the surface 0.75 m, and the greatest proportion of the soybean root system occurred near the soil surface (13). The considerably greater amount of water removed from deeper in the soil by tall morning glory, especially after 6 weeks after planting can be attributed to a greater exploration of the soil volume by its roots (11). This ability to gain access to available water explains one of the mechanisms by which tall morning glory is competitive with agronomic crops (14). Under interspecific plant conditions the amount

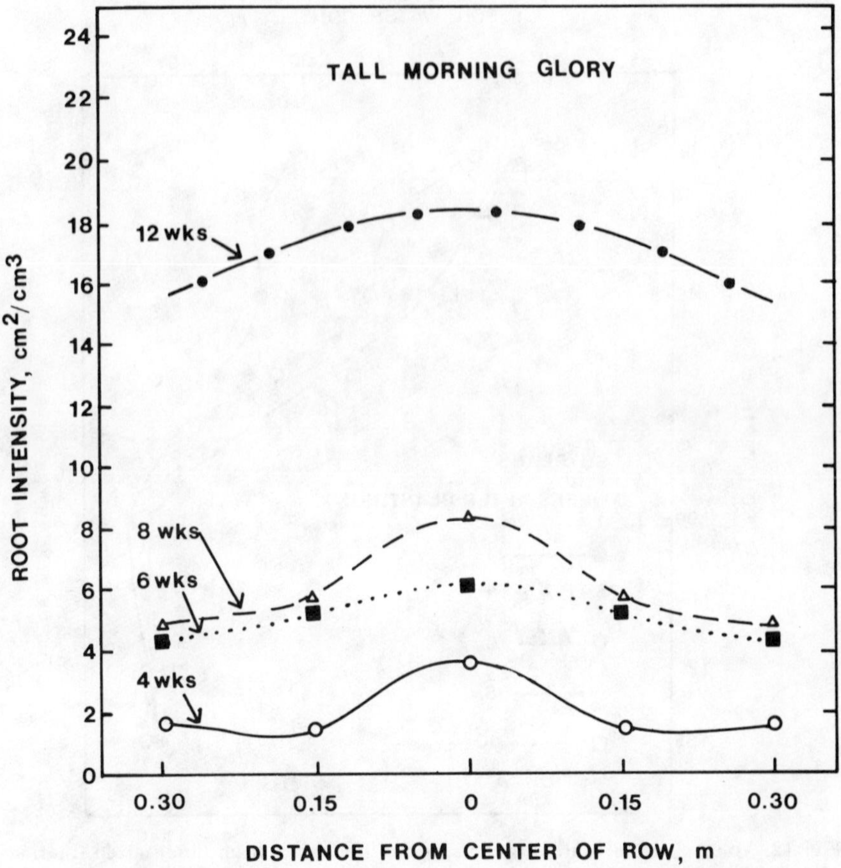

Fig. 11. Relationships between tall morning glory root intensity and distance from row center at selected plant ages. The tall morning glory root density was 6.6 plants/m.

of water extracted from the soil profile was similar to that extracted by soybeans alone during the early part of the growing season and to that by tall morning glory during the later part of the season. Since the extraction effects were not additive, this result indicates that the soil and climate placed restrictions on the amount of water extracted. Similar competition results were found by Geddes et al. (28) with another aggressive weed, common cocklebur (*Xanthium pensylvanicum* Wallr.), and soybeans. The density of roots of both species was greatest in the 0 to 13 cm depth interval and in the dry season more roots were found deeper in the soil. Roots of common cocklebur and the interspecific treatment explored a greater volume of soil than did those of soybeans alone. The amount of water used by the interspecific soybean-common cocklebur was greater than that of the intraspecific cocklebur.

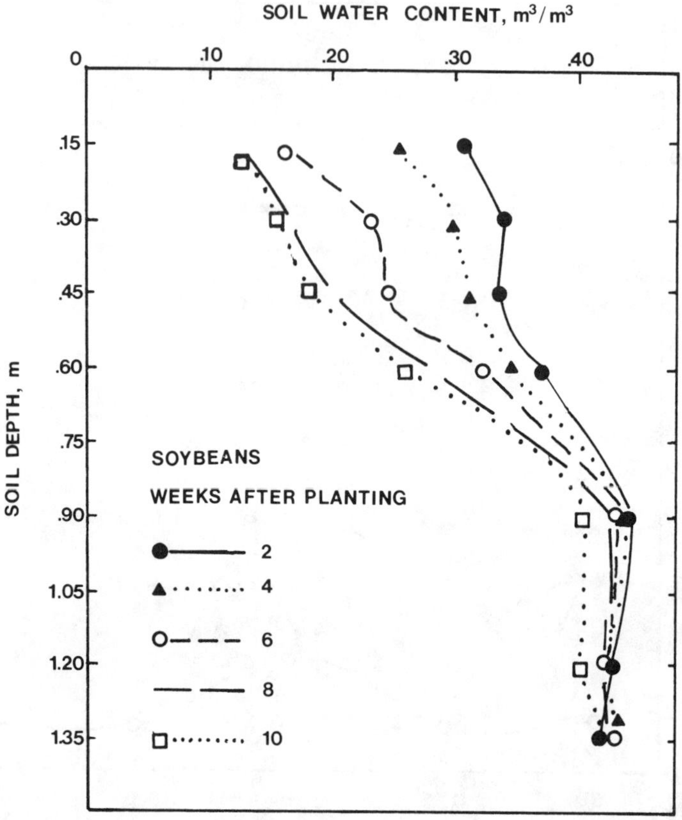

Fig. 12. Volumetric soil water contents in the Taloka silt loam (fine, mixed, thermic Mollic Albaqualf) profile of the interspecific soybeans at selected plant ages.

Biological Stresses

Shoot and Root Stress. Distribution of roots in the soil can be described as the concentration of roots in the x, y, and z planes of the soil profile (Fig. 3). Root concentration usually has been defined in terms of root weight, length, or surface area per unit soil weight or volume. The concentration of roots in any one of these planes can change abruptly with soil and climatic conditions during vegetative and reproductive growth. This abrupt change in the concentration of roots in the soil may be due to biological stress conditions in either the shoot or the root system which results in a failure of the plant to maintain its normal physiological functions required for uninterrupted growth and development.

A biological stress occurs when an external stimulus acts on a portion of the shoots or the root system to the extent that the normal growth is significantly reduced, modified, or stopped.

The seasonal pattern of dry matter accumulation and nutrient and water absorption by soybeans under nonstressed conditions represents

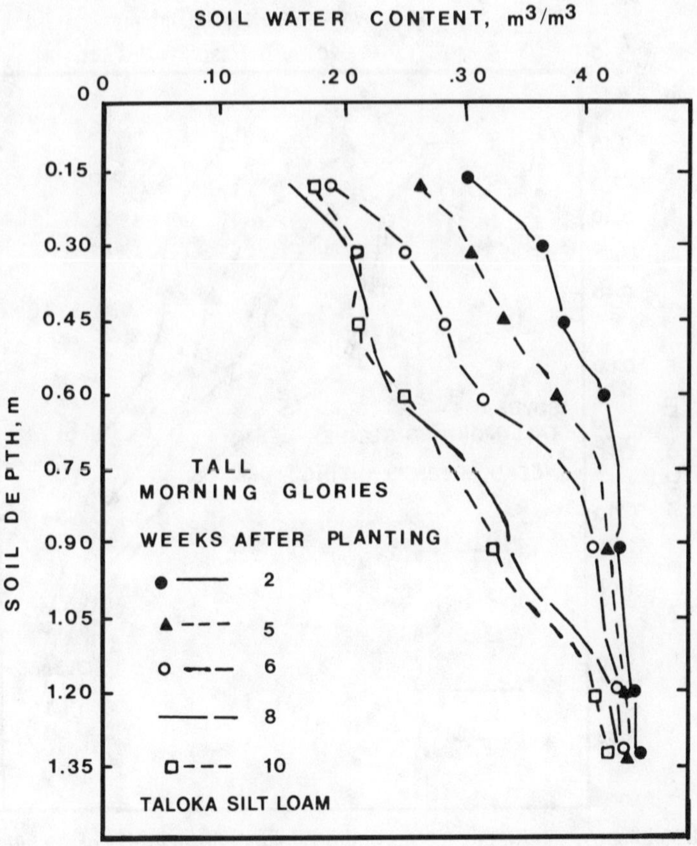

Fig. 13. Volumetric soil water contents in the Taloka silt loam profile of the intraspecific tall morning glories at selected plant ages. The tall morning glory planting density was 6.6 plants/m.

an optimum growth rate of the root and shoot (Fig. 15). The interruption of the growth pattern of either the shoot or root by a significant stress condition is reflected in abrupt changes in root distribution in any one or each of the directional planes (x, y, z) within the soil profile.

It can be concluded that root distribution in the soil profile is, therefore, a function of the interaction of the genetic characteristics of the plant with the microclimate in the plant canopy and the physical, chemical, and microbiological properties of the soil profile. Under favorable growth conditions the environment allows the plant to express its genetic capabilities more fully for shoot and root growth throughout vegetative and reproductive growth. Under such conditions, the plant should more nearly achieve its potential yield.

The failure of a plant to reach its potential yield will, therefore, be due to stress conditions of varying degrees of severity and duration developing either in the shoot or root system of the plant. Since there

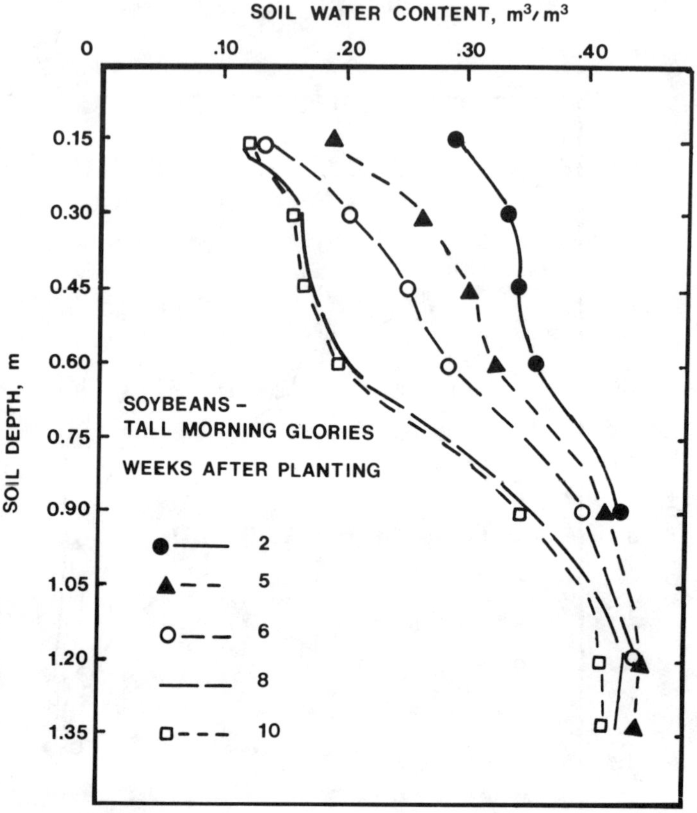

Fig. 14. Volumetric soil water contents in the Taloka silt loam profile of the interspecific soybeans-tall morning glories at selected plant ages. The tall morning glory planting density was 6.6 plants/m.

is a functional equilibrium between the growth of the shoot and root, a significant stress in either the canopy or the root system induces responses throughout the plant.

Stress in the shoot resulting from excessive temperature and evaporative demand or inadequate light which reduces photosynthetic activity not only slows shoot growth but also limits photosynthate and assimilate translocation to the roots, thereby slowing the rate of root growth. The decline in root growth subsequently decreases the flow of water and nutrients to the shoot. Thus, a stress originating in the canopy is transmitted at varying degrees of severity throughout the entire plant.

In a similar manner, when rapidly growing roots interact with a soil horizon having a reduced availability of water, air, or nutrients, the terminal growth of the roots may decrease or completely cease. Thus, the supply of water and nutrient translocated to the shoot can be severely limited. As a result, the shoot may slow its rate of terminal growth functions—including photosynthesis and assimilation, leaf expansion, and flower initiation or retention (31, 34).

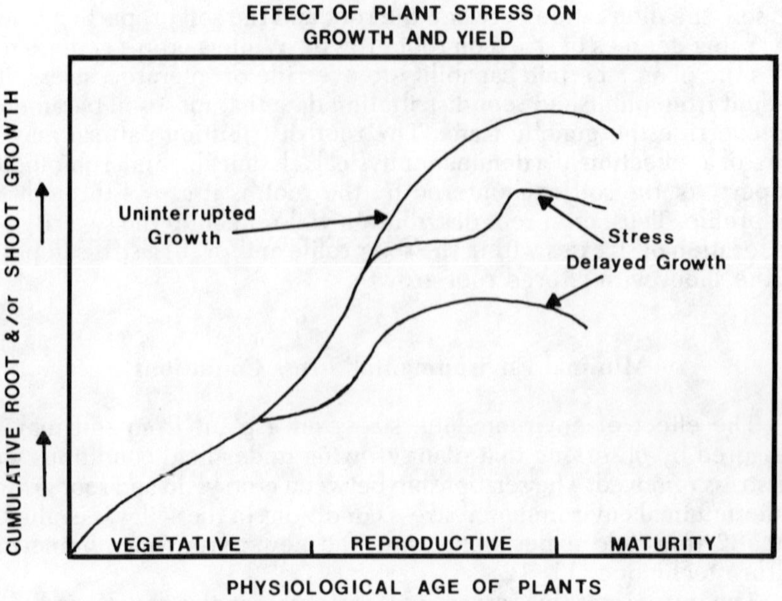

Fig. 15. Relationship between cumulative root and shoot growth as affected by stress conditions and plant age.

An additional reaction by the plant to severe stress conditions involves the resumption of the normal growth rate after the stress has been removed. When terminal root growth has been severely limited, there may be as much as an 8-day delay in the resumption of new root growth (Fig. 15). The cessation of terminal root growth generally results in new lateral branches induced from the older portions of the root system into regions of the soil that have more favorable growing conditions in the profile.

The delay in new root growth, which results in decreased translocation of nutrients and water to the shoot, imparts a significant stress in the shoot (5, 26). During this stress period and before resumption of normal growth, the shoot may not set those flowers which it would otherwise set under nonstressed conditions; flowers set at the onset of the stress may not be retained; or young pods of soybeans or squares of cotton plants may be dropped. In effect, as much as an 8-day growth period may occur in which the plant will not set new flowers or hold young fruit, leaving an unfruited area within the canopy. The absence of a bottom crop, or a middle crop in soybeans or cotton is a frequently observed phenomenon in crop growth under unfavorable growing conditions. Under such growth conditions the plant cannot approach its potential yield.

The genetic characteristics of a given cultivar growing under nonstressed conditions develop a pattern of root development specific to that cultivar. However, under field conditions, as the roots grow through the soil, the interaction between the root and the soil properties results in varying degrees of stress on root growth. While a crop's genetic trait gives the plant a certain capability to override or tolerate a stress, it is evident from published root distribution data that most soil parameters can override the genetic traits. The root distribution pattern may be more of a reflection of a dominant physical, chemical, or microbiological property of the soil encountered by the root as it grows through the soil profile. Thus, crop root distribution will reflect (a) the severity and the location of stresses within the soil profile and/or (b) a stress imposed on the shoot with altered root growth.

Minimal Environmental Stress Conditions

The effect of environmental stress on a plant from soil may be measured by observing that plant growing under field conditions with the stress removed. The relationship between crop yield and root growth under minimal environmental stress conditions in the soil was evaluated in 1982 in a field experiment using the new porous membrane-root culture technique (11, 12).

The purpose of this research was to measure the growth and yield of the soybean when its roots were allowed to grow uninhibited within a porous membrane envelope under favorable conditions for air, nutrients, and water. Under these conditions, the plant obtained all of its nutrients, except N, from the soil solution. The plant roots were permitted to grow under minimal stress from unfavorable soil properties which allowed them to approach their full growth potential. The plant growth parameters for the soybean cvs. R80 to 84, are given in Table 2.

The excellent growth of the plants throughout the season proved conclusively that when a soybean plant is allowed to grow under minimal stress conditions in the field it is capable of growing to a height of 1.08 m, produce a leaf area of 1.9 m^2, and set from 600 to 1050 pods per plant. The maximum root extension in the membrane was 2.5

Table 2. Changes in the characteristics of soybean plants grown under minimal stress conditions in the field. (Porous membrane root culture technique) (12).

	Growth stage					
Plant characteristic	V-8	V-12	R-1	R-3	R-5	R-8
Plant height (cm)	38	52	56	67	104	108
Leaf area (cm^2)	1 909	3 251	3 396	8 245	19 593	--
Root extension (cm)†	79	89	102	102	102	192
Shoot dry matter (g)	11.2	27.0	35.0	83.2	273.3	--
Root dry matter (g)	5.8	12.0	14.9	20.8	21.5	35.8
Shoot:Root ratio	1.93	2.24	2.34	4.00	11.00	--
No. pods/plant‡	--	--	--	100	577	846

† Maximum length in membrane in the R-1, R-3, and R-5 root growth was limited by membrane length; plants grown in the membrane with extended length (240 cm) was 210 cm.

‡ Included 51% 3-seeded pods; 29% 2-seeded pods; 14% 1-seeded pods; and 6% pods with no seed.

m in the longest membrane (Fig. 16 and 17). Excellent nodule development occurred throughout the entire 2.10-m root length (Fig. 18). Nitrogen supplied by N_2-fixing bacteria was adequate for plant growth (3.5% in leaves).

When grown under high plant populations (26 plants/m), plants of this same cultivar produced roots in the soil profile to a depth of 0.79 m, an average leaf area of 0.25 m^2, and 98 pods per plant. When grown

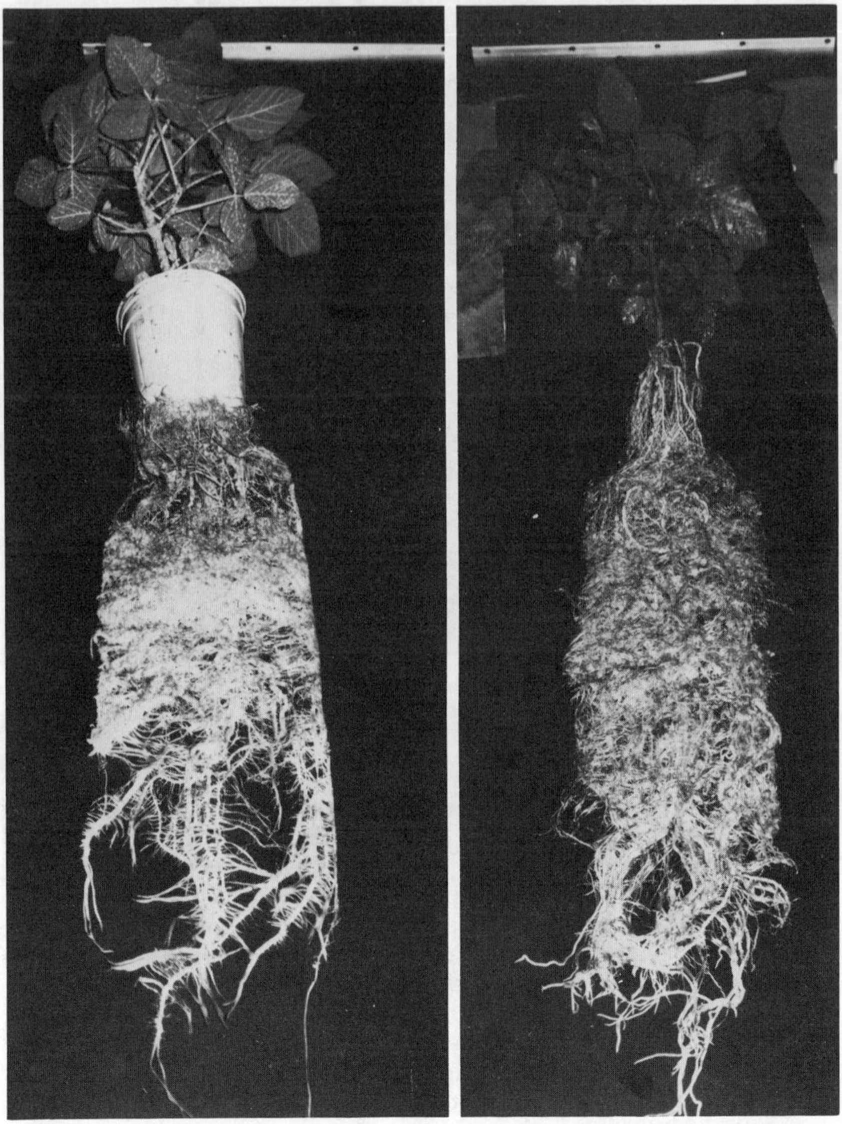

Fig. 16. Soybean root growth via the porous membrane-root culture technique (11). Root growth at three stages of plant maturity; Left. V8 (0.5 m), Right. R5 (2.5 m). Length expressed as distance from the crown of the plant.

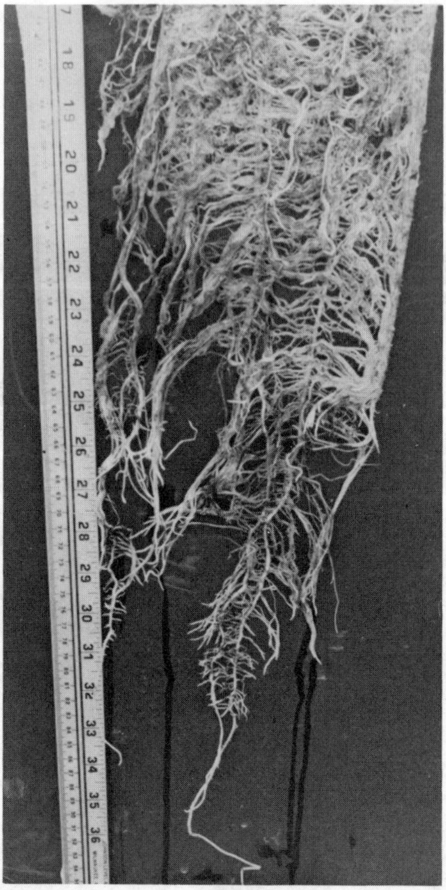

Fig. 17. A. Close-up view of the terminal portion of soybean roots after removal from the porous membrane at the R5 stage of growth.

as a single plant and fertilized and watered at the same rate as those growing in the membrane envelope, plants of this cultivar produced approximately one-third fewer pods (508). The data show the difference between 846 and 98 seed pods per plant was due to stress on the shoot portion of the plant created by the root competition for soil water and nutrients plus the effect of unfavorable physical conditions in the soil upon the rate of root growth and distribution.

Another significant part of these data is the increase in the number of seed pods set during plant maturity. The plant had 100 pods at the R1 stage, 576 at the R3 stage, and 846 at the R8 stage. Thus, the plant set and matured 32% of its total pods during the late reproductive stages (R5 to R8). This finding supports the concept that the continuous growth of new roots is essential in the late reproductive stage for the plant to approach its potential yield.

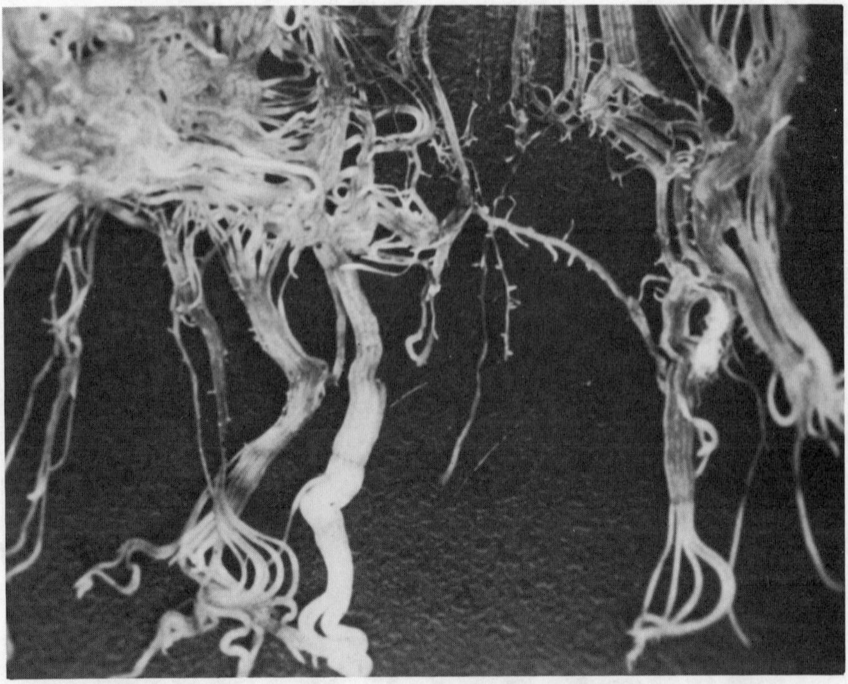

Fig. 17. B. Terminal roots showing the profuseness of new root initiation at the R5 stage of growth.

ROOT FLUORESCENCE

Another significant root parameter which can be measured by the porous membrane-root culture technique is defined as the fraction of the total root system which is actively elongating as indicated by new root extension. Recent research by Dyer (21) and Dyer and Brown (23) has shown that the most actively growing portion of soybean root fluoresces strongly under ultraviolet (UV) light (365 nm wavelength). A positive correlation was found between the intensity of fluorescence and rate of new root growth and ion absorption (Fig. 19). The fluorescent intensity was found to decrease with root age and to disappear when suberization began (Fig. 20). This relationship has been used to calculate either the amount of roots fluorescing or the fraction of the total root system fluorescing as a measure of the root activity at different stages of plant maturity. This calculation is possible from the length (or surface area) of roots as determined from photographs under white light (total root) and then under fluorescent light (active elongating portion of the root system).

Soybean plants grown via the porous membrane-root culture technique in 1982 provided excellent root systems for photographing under white and UV light. The root systems for six stages of growth were photographed in 15-cm sections throughout the entire root length for

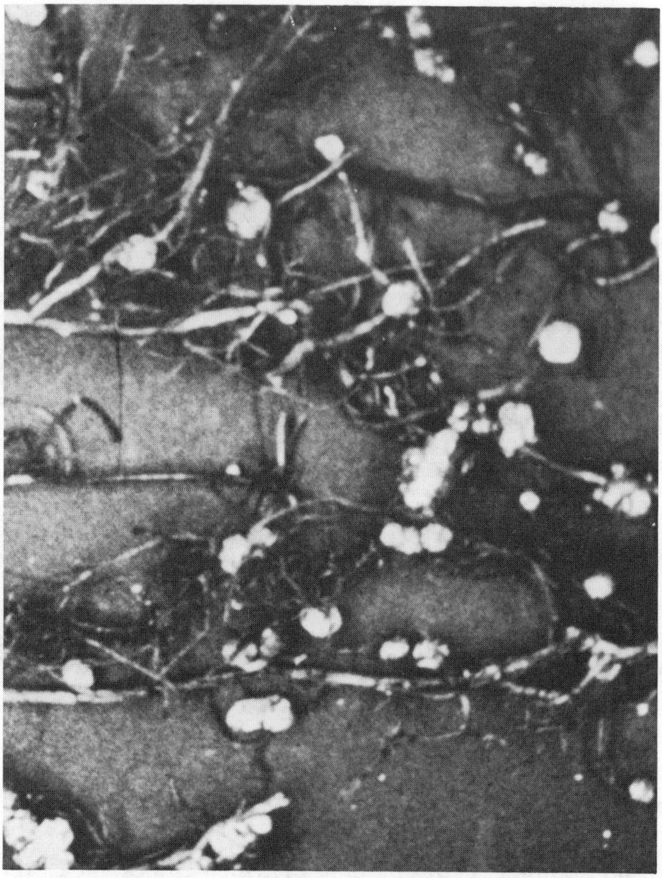

Fig. 18. Soybean roots exhibited excellent nodule formation throughout its root; photograph shows nodulation at 1.4 m from crown.

six growth stages. The fraction of the total roots fluorescing at two positions within the root system for each stage of growth are shown in Fig. 21. The fraction of the root system fluorescing in the root-tip section (0 to 0.15 m) decreased from 24% at the V8 stage of growth to 1.4 at the R6 stage. In the older roots (i.e., in the center of the root system 15 to 21 cm from root tip), there was only 4.2% fluorescing at the V8 stage, and this rate decreased to zero at the R6 stage. The largest amount of fluorescent roots was always present in the terminal (most recent growth) portion of the root system (Fig. 20); the least was in the oldest portion of the root system. The percent of fluorescent roots decreased from 27.9 to 4.2 in the V8 stage, while the decrease was from 3.5 to 1.0 in the R5 stage.

The change in active root elongation for the terminal portion of the root decrased exponentially from the V8 to the R3 stage of growth. The continued rapid decrease between V12 and R3 stages was primarily

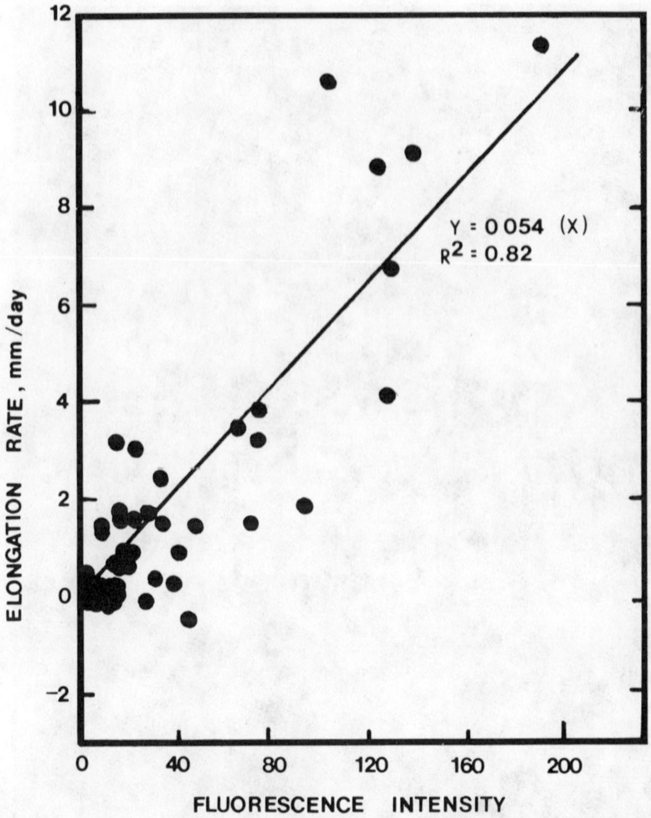

Fig. 19. Relationship between the rate of soybean elongation and fluorescent intensity under ultraviolet light (365 nm). Dyer and Brown (23).

due to the fact that the roots reached the end of the membrane envelope which slowed their rate of new growth. While this represents a decrease in the percentage of fluorescing roots of the total roots present, the total amount of fluorescent root growth actually increased from the V8 to the R5 stage.

The percentage of fluorescing roots in the central portion of the root system was about sevenfold less than that of the terminal roots at the V8 stage of growth. The percentage of fluorescing roots remained constant to the R3 stage, when the growth decreased suddenly to 0.6 and to 0.0 for the R3 and R8 stage of growth, respectively.

The data suggest that the amount of the total root system which is actively growing and elongating may be more valuable as an indicator of how well the plant is performing than the use of other root parameters. Plants undergoing physiological stress in the shoot or root will show a decrease in new root growth. Since the amount of new root growth is a prime factor in water and nutrient absorption, this parameter should correlate better with plant yield than other conventional parameters such as root mass, area, size, or length.

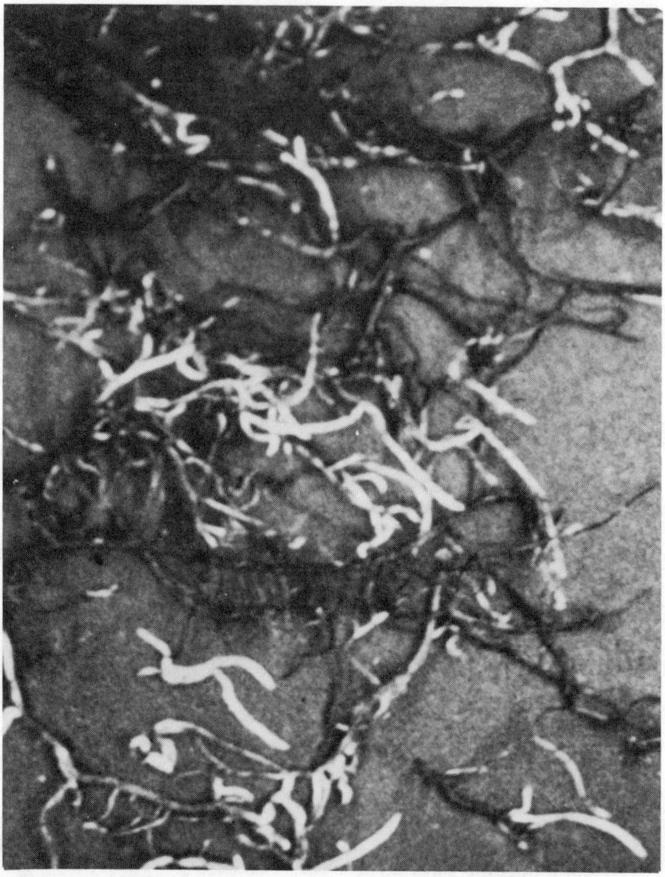

Fig. 20. Soybean roots photographed under ultraviolet light. Most recent roots fluoresce and appear white; the older nonfluorescing roots appear dark in color.

In this experiment, the porous membrane root culture technique prevented the soybean roots from accumulating gelatinous organic and inorganic coatings on the surfaces of the older root surfaces. In this case, the older roots of these plants gave evidence of remaining active in water and nutrient absorption during the late reproductive stage of growth. This condition does not appear to be true for soybean roots growing freely in the soil profile. In this situation, the older roots accumulate thick, gelatinous coatings which contain an assortment of exudate and organic and inorganic materials which have the capacity to decrease the rate of water and nutrient absorption. As a result, the new developing roots become the primary region of uptake during the latter stages of growth.

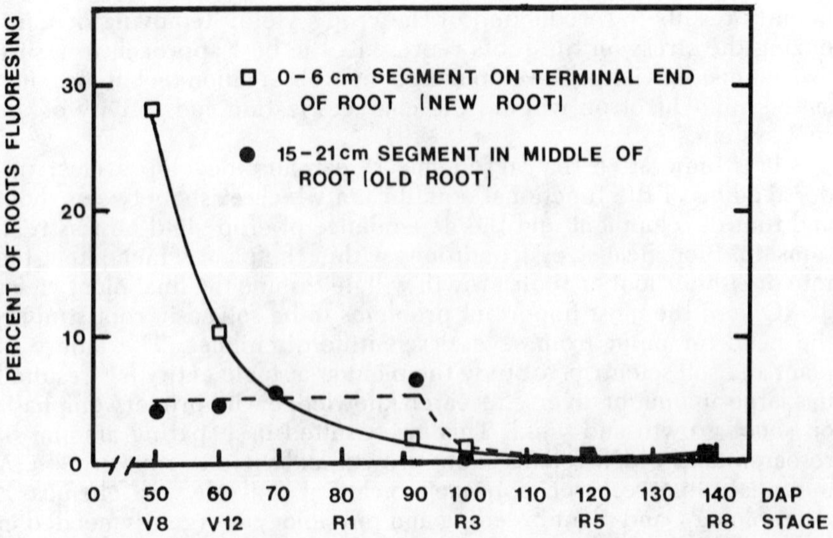

Fig. 21. Changes in the percentage of soybean roots elongating at two positions in the root system for six stages of growth. Cultivar R84–80. Brown and Haq (13).

CONCLUSIONS

The distribution of roots in the soil results from a series of complex and dynamic processes which include the interaction of environment, soil, and the actively growing plants. Since water and nutrients are supplied by the root system to the shoot, it is the concentration, distribution, and activity (growth rate and longevity) of the roots within the soil profile which will largely determine crop yield. Roots, which have their growth rate slowed or interrupted due to internal physiological stress arising from either the soil or the shoot-microclimate, will systematically reduce the capacity of the plant to reach its potential yield.

In order to relate crop yields to root growth and distribution, the data must be interpreted in terms of (a) age of the plant; (b) microclimate conditions in the plant canopy; and (c) the physical, chemical, and microbiological conditions throughout the soil profile.

The most frequent soil conditions which determine the root growth and distribution pattern include: (1) depth of root zone; (2) characteristics of the root zone which determine soil density, aeration, water movement, nutrients, pH, and presence of toxic elements; and (3) the degree of intra- or interspecific root competition resulting from high plant population or infestation of weeds.

Absorption activity or growth rates slowed or stopped due to internal physiological stress from the root and/or shoot environment frequently results in a reduction of the crop's yield. Removing or minimizing the stress on the roots represents the best approach to assure maximum crop yields. Development of new observation techniques such as the mini-rhizotron will aid the characterization and activity of the root system.

It is imperative that agronomic researchers develop a clear understanding of the functional equilibrium which exists between shoot and root development and the dependence of crop yield on this relationship. Biological stress conditions within the plant which affect the rate of either shoot or roots growth will determine the final plant yield.

One of the most important problems to be solved in root study is the need for input from several scientific disciplines. The failure of plant and soil scientists to study the plant as a single entity has resulted in a large amount of diverse research knowledge with primary emphasis on shoot growth and yield. This has resulted in a sparing amount of research data which relates shoot growth and yield to root growth. A multidiscipline research approach including soil physics, chemistry, microbiology, and plant breeding and physiology is urgently needed in order to establish the significance of the plant root system in plant performance and yield. Multidiscipline research represents a wholistic approach which involves consideration of the entire plant as a whole entity.

REFERENCES

1. Adams, F., R.W. Pearson, and B.D. Doss. 1967. Relative effects of acid subsoils on cotton yields in field experiments and on cotton roots in growth chamber experiments. Agron. J. 59:453–456.
2. Agrawal, R.P., R.K. Khanna, J. Nath, and M.L. Barta. 1975. Root penetration studies with 32P in cereals as affected by compact layers at varying depths. Ann. Arid Zone 14:339–346.
3. Allmaras, R.R., and W.W. Nelson. 1973. Corn root-configuration response to soil temperature and matric suction. Agron. J. 65:725–730.
4. ----, ----, and W.B. Voorhees. 1975. Soybean and corn rooting in southwestern Minnesota: II. Root distributions and related water inflow. Soil Sci. Soc. Am. Proc. 39:771–777.
5. Arkin, G.F., and H.M. Taylor. 1981. Modifying the root environment to reduce crop stress. Am. Soc. of Agric. Eng., St. Joseph, MI.
6. Barber, S.A. 1971. Effect of tillage practice on corn (*Zea mays* L.) root distribution and morphology. Agron. J. 63:721–726.
7. Bassett, D.M., J.R. Stockton, and W.L. Dickens. 1970. Root growth of cotton as measured by P32 uptake. Agron. J. 62:200–203.
8. Böhm, W. 1974. Mini-rhizotrons for root observation under field conditions. I. Arch. Acker Pflanzenbau Bodenkd. 140:282–287.
9. ----. 1979. Methods of studying root systems. Springer-Verlag, New York.
10. Brouwer, R. 1963. Some aspects of the equilibrium between overground-underground plant parts. Jaarb. I.B.S. (Wageningen) 31–39.
11. Brown, E.A., C.E. Caviness, and D.A. Brown. 1980. Evaluation of selected soybean cultivars for tolerance to drought stress. Progress Rep. to Am. Soybean Assoc. Agronomy Dep., Univ. of Arkansas, Fayetteville, AR.

12. Brown, D.A., and Anwar Haq. 1983. Physiological stress on roots: A barrier to maximum soybean yields. Arkansas Farm Res. 32:9.

13. ----, and ----. 1984. A porous membrane-root culture technique for growing plants under controlled soil conditions. Soil Sci. Soc. Am. J. 48:692–695.

14. Burgess, T.M., and R. Webster. 1980a. Optimal interpolation and isarithmic mapping of soil properties. I. The semi-variogram and punctual kriging. J. Soil Sci. 31:315–331.

15. ----, and ----. 1980b. Optimal interpolation and isarithmic mapping of soil properties. II. Block kriging. J. Soil Sci. 31:333–341.

16. Cappy, J.J. 1979. The rooting patterns of soybean and cotton throughout the growing season. Ph.D. Diss. Univ. of Arkansas. Abstr. 40, 1979. Pt. 2. p. 995–1968B (Pub. no. 7919267).

17. Carithers, Clay. 1981. Corn rooting pattern and yield as affected by a growth regulator. Master's thesis. Univ. of Arkansas, Fayetteville, AR.

18. Clarkson, D.T., and G.C. Gerloff. 1979. The effects of temperature on root growth and mineral nutrition of young maize plants. p. 52–54. *In* Annual Rep. to the Agric. Res. Coun. Letcombe Lab., Wantage, Oxon, England.

19. Curry, R.B., C.H. Baker, and J.G. Streeter. 1975. SOYMOD I: A dynamic simulator of soybean growth and development. Trans. ASAE 18:963–968, 974.

20. Das, K.C. 1973. Dynamics of corn root growth as affected by compact subsoil and its influence on crop response to irrigation. Diss. Abstr. Int. B:35(4):1338.

21. Dyer, D.J. 1980. Initial investigations of root fluorescence in soybean: cause and applications. Master's thesis, Univ. of Arkansas, Fayetteville, AR.

22. ----. 1982. Interpretive and predictive modeling of crop root systems. Ph.D. Diss. Univ. of Arkansas. (Diss. Abstr. vol. 43, no. 10, p. 3080B–3419B, April 1983; Pub. no. DA8305052).

23. ----, and D.A. Brown. 1983. Relationship of fluorescent intensity to ion uptake and elongation rates of soybean roots. Plant Soil 72:127–134.

24. Eavis, B.W. 1972. Soil physical conditions affecting seedling root growth III. Comparisons between root growth in poorly aerated soil and at different oxygen partial pressures. Plant Soil 37:151–158.

25. Fehr, W.R., and C.E. Caviness. 1977. Stages of soybean development. Spec. Rep. 80. Iowa Station. Iowa State University.

26. Fritter, H.H., and R.K.M. Hay. 1981. Environmental physiology of plants. Academic Press, New York.

27. Gajem, Y.M., A.W. Warrick, and D.E. Myers. 1981. Spatial dependence of physical properties of a typic torrifluvent soil. Soil Sci. Soc. Am. J. 95:709–715.

28. Geddes, R.D., H.D. Scott, and L.R. Oliver. 1979. Growth and water use by common cocklebur (*Xanthium pensylvanicum*) and soybeans (*Glycine max*) under field conditions. Weed Sci. 27:206–212.

29. Hajrasuliha, S., N. Baniabbassi, J. Metthey, and D.R. Nielsen. 1980. Spatial variability of soil sampling for salinity studies in southwest Iran. Irrig. Sci. 1:197–208.

30. Halevy, J. 1976. Growth rate and nutrient uptake of two cotton cultivars grown under irrigation. Agron. J. 68:701–705.

31. Howell, R.W. 1982. Wither soybeans physiology. Soybean News. Vol. 34, No. 1.

32. Huck, M.G. 1970. Variation in taproot elongation rate as influenced by composition of the soil air. Agron. J. 62:815–818.

33. Kasper, T.C., C.D. Stanley, and H.M. Taylor. 1978. Soybean root growth during the reproductive stages of development. Agron. J. 70:1105–1107.

34. Levitt, J. 1980. Responses of plants to environmental stresses. Academic Press, New York.

35. Matheron, G. 1963. Principles of geostatistics. Econ. Geol. 58:1246–1266.

36. Mayaki, W.C., L.R. Stone, and T.D. Teare. 1976. Irrigated and nonirrigated soybean, corn, and grain sorghum root systems. Agron. J. 68:532–534.

37. Mengel, D.B.; and S.A. Barber. 1974. Development and distribution of the corn root system under field conditions. Agron. J. 66:341–344.
38. Mitchell, R.L., and W.J. Russell. 1971. Root development and rooting patterns of soybean (*Glycine max* (L.) Merrill) evaluated under field conditions. Agron. J. 63:313–316.
39. Narda, J.K., and R.B. Curry. 1981. *SOYROOT*—A dynamic model of soybean root-growth and water uptake. Trans. ASAE 24:651.
40. Nielsen, D.R., J. Metthey, and J.W. Biggar. 1981. Soil hydraulic properties, spatial variability, and soil-water movement. p. 47–68. *In* I.K. Iskandar (ed.) Modeling wastewater renovation: Land treatment. John Wiley & Sons, New York.
41. Onderdonk, J.J., and J.W. Ketchenson. 1973. Effect of soil temperature on direction of corn root growth. Plant Soil 39(1):177–186.
42. Page, E.R., and A. Gerwitz. 1974. Mathematical models, based on diffusion equations, to describe root systems of isolated plants, row crops, and swards. Plant Soil 41:243–254.
43. Papendick, R.I., and J.R. Runckles. 1965. Transidentstate oxygen in soil: I. The case when rate of oxygen consumption is constant. Soil Sci. 100:251–261.
44. ----, and ----. 1966. Transidentstate oxygen diffusion in soil: II. A case when rate of oxygen consumption varies with time. Soil Sci. 102:223–230.
45. Pearson, R.W. 1965. Soil environment and root development. p. 95–126. *In* W.H. Pierre et al. (ed.) Plant environment and efficient water use. Am. Soc. of Agron. and Soil Sci. Soc. of Am., Madison, Wis.
46. Raper, C.D., and S.A. Barber. 1970. Rooting systems of soybean. I. Differences in root morphology and varieties. Agron. J. 62:581–584.
47. Robertson, W.K., L.C. Hammond, J.T. Johnson, and K.J. Boote. 1980. Effects of plant-water stress on root distribution of corn, soybeans, and peanuts in a sandy soil. Agron. J. 72:548–550.
48. Sanders, J.L., and D.A. Brown. 1976. The effects of variations in the shoot-root ratio upon the chemical composition and growth of soybeans. Agron. J. 68:713–716.
49. ----, and ----. 1978. A new fiberoptic technique for measuring root growth of soybeans under field conditions. Agron. J. 70:1073–1076.
50. ----, and ----. 1979. Measurement of rooting patterns for determinate and indeterminate soybean genotypes with a fiber-optic scope. p. 369–379. *In* Proc. of the Int. Cont. on Soil-Root Interface. Oxford, England.
51. Schuurman, J.J. 1965. Influence of soil density on root development and growth of oats. Plant Soil 22:352–374.
52. Scott, H.D., and L.R. Oliver. 1976. Field competition between tall morning glory and soybean. II. Development and distribution of root systems. Weed Sci. 24:454–460.
53. ----, R.E. Sojka, D.L. Karlen, F.B. Arnold, V.L. Quisenberry, and C.W. Doty. 1983. Bragg soybeans grown on a southern coastal plains soil. I. Dry matter distribution, nodal growth analysis, and sample variability. J. Plant Nutr. 6:133–162.
54. Sharp, R.E., and W.J. Davies. 1979. Solute regulation and growth by roots and shoots of water-stressed maize plants. Planta 147:43–49.
55. Taylor, H.M., and B. Klepper. 1973. Rooting density and water extraction patterns for corn (*Zea mays* L.). Agron. J. 65:965–968.
56. ----. 1971. The Rhizotron at Auburn, AL: Design and three years use. *In* Proc. 3rd Int. Seminar for Hydrology Professors. June. Purdue University, Lafayette, IN.
57. Vierra, S.R., D.R. Nielson, and J.W. Bigger. 1981. Spatial variability of field measured infiltration rate. Soil Sci. Am. J. 45:1040–1047.
58. Walker, J.N., and W.E. Splinter. 1971. Introduction: Mathematical modeling of plants. Trans. ASAE 14:945.
59. Webster, R., and T.M. Burgess. 1980. Optimal interpolation and isarithmic mapping of soil properties. III. Changing drift and universal kriging. J. Soil Sci. 31:505–524.
60. West, W.S. 1976. The effect of shoot-root ratio upon the nutrient uptake and yield of cotton. Master's thesis, Agronomy Dep., Univ. of Arkansas, Fayetteville, AR.